ARBORETUM AMERICA

ARBORETUM AMERICA

A Philosophy of the Forest

Diana Beresford-Kroeger

Photographs by Christian H. Kroeger • Foreword by Edward O. Wilson

The University of Michigan Press
Ann Arbor

Published in the United States of America by
The University of Michigan Press
Manufactured in Canada
♾ Printed on acid-free paper

2006 2005 2004 2003 4 3 2 1

A CIP catalog record for this book is available from the British Library.

Library of Congress Cataloging-in-Publication Data
Beresford-Kroeger, Diana, 1944–
Arboretum America : a philosophy of the forest / Diana Beresford-Kroeger ; photographs by Christian H. Kroeger ; foreword by Edward O. Wilson.
p. cm.
Includes bibliographical references (p.)
ISBN 0-472-09851-9 (cloth : alk. paper) — ISBN 0-472-06851-2 (paper : alk. paper)
1. Trees—North America. I. Title.
QK110.B475 2003
582.16'097—dc21 2003010310

Frontispiece: A young woodland of sugar maple, *Acer saccharum,* in spring

THIS BOOK IS DEDICATED TO

Patrick O'Donoghue, a maternal, bachelor uncle, under whose wardship I was placed, at a young age, upon the death of my family. He insisted that knowledge was no burden and that erudition in a woman was more important than in a man.

Iwan Kroeger, my father-in-law. His dream on the shores of the Baltic put man on the moon in the Apollo space system. He followed the path of truth in science. He once told me that I was the daughter he never had. He believed in me and in his son, Christian.

Aralia nudicaulis, wild sarsaparilla, is a member of the ginseng family. It thrives in the dappled shade of the forest floor.

Acknowledgments

For their generosity in allowing us full range of their pristine private forest, my husband, Christian, and I wish to thank Mr. and Mrs. Carl Campitelli. Dr. E. Small and Ms. Gisèle Mitrow gave much support in opening the Canadian Vascular Plant Collection at the Eastern Cereal and Oilseed Research Centre in Ottawa. We also want to thank Mr. Brian Douglas of the Dominion Arboretum in Ottawa for his support. Mrs. Nancy Wortman typed this manuscript, offering advice and cups of tea throughout. The technical support by her husband, Lynn, was much appreciated. We affectionately wish to thank my daughter, Erika, for her patience as we ploughed through this manuscript in her final school year. We acknowledge Ms. Mary Kate Laphen, of our local library, the Merrickville Public Library, for her cheerful help.

Finally I wish to say that without the continuing support of my husband and his editorial skills this book would never have seen the light of day.

Precambrian rock with the frosted rock tripe, *Umbilicaria americana*, the edible lichen. This is the real beginning of a forest.

Foreword

EDWARD O. WILSON

Who speaks for the trees, speaks for all of nature. Diana Beresford-Kroeger is one of the rare individuals who can accomplish this outwardly simple but inwardly complex and difficult translation from the nonhuman to human realms. She does so from the kind of intimate knowledge and love of a subject that can be acquired only by a lifetime of experience. Poet and naturalist by calling, she is both druidical and scientific in literary expression. Her portraits of common North American trees are blended with uncommon skill from botany, ethnology, silviculture, and history, but they can be admired for the felicity of the language alone.

The result of her effort is to imbue each Nearctic tree species addressed with a personality. Her portraits are not, however, anthropomorphic. They are solidly based on fact, and acquire their richness and value from the multiple dimensions of the information provided. Each species, we are reminded from the examples she provides, is a masterpiece of evolution, exquisitely adapted to the environment in which it has evolved for millions of years.

Beresford-Kroeger and I share a dream. We want people to see the forest *and* the trees, and the wildlife abounding in wild environments, in fine detail. We want native species to be valued and cultivated one by one for the special place they have in the deep history of the land. We want horticulturists to contribute more consciously and joyfully to the precious natural heritage they represent. *Arboretum America* speaks for the trees as well as it has ever been done.

Catalpa speciosa in full bloom in July

Contents

A holy place

Introduction

Comfort food came to me from a tree. The tree was a sweet chestnut, *Castanea sativa*. I was around six or seven years old at the time and had to climb a two-hundred-foot rock face behind my home. The top of this cliff extended into a little valley. My tree grew in luxurious abandon, throwing out limbs in a strong, horizontal manner and aiming its head toward an Irish sky. In the autumn I would sit on one of these branches midway up the tree and eat prodigious amounts of sweet-fleshed nuts. Then I would swing higher to look out over the gray of the slate rooftops declining toward the River Lee. I would watch the boats, packed into Cork harbor, testing their moorings as the river pushed its way through the estuary to the Atlantic.

It was some years later when I was caged in the uniform of a private girl's school and my mother had decided to open my mind to the delights of the theater in London that I saw chestnuts again. We had come out into the slippery darkness of a rain-filled London night after the joys of the stage and its bright lights. My mother spotted a vendor roasting nuts. She sent somebody over to this soaking man for a bag of roasted chestnuts for me. On our way back to Gloucester Road in Kensington the feel of my wild tree in Ireland came rushing into my mind. I commented casually to all present, "Why these nuts are *completely* different." My mother answered, "But, my darling, wherever have you tasted chestnuts before?" I continued to eat in silence. For silence is golden and mothers never, ever know what their children get up to.

To me, the world is a garden, a global garden. Like all gardens, it is finite in its ability to grow trees. We do not realize that if we keep cutting, we will not have anything left. I still feel like that child perched on the sweet chestnut tree; I think we can replant the global forests. This book marks a beginning. To begin is to succeed.

I believe in the more natural approaches to the care of our own gardens and that of the global garden. This started for me many years ago when I was doing research. I modified the electron microscope to capture a phenomenon I have called cathodoluminescence. It is a form of light given off by special, excited electrons in a carbon ring structure. By chance I decided to examine a common pesticide. Although I wore protective gloves, the chemical caused an extraordinary, weeping burn on my body for months. The thought that went through my head again and again was, "And they put that on food crops. Dear God!"

I have always been fascinated by the medicines of nature. If a tree can live for a thousand years, then it has to have an excellent internal system of protective medicine. These medicines have been used by mankind since time immemorial. There are more medicines waiting to be discovered in trees. These are the volatile fragrances and other chemicals that move on airways around trees and in gardens. These chemicals are the adaptive molecules of evolution: they hold the "smart card" for the human race.

Because the water in the global garden is becoming increasingly contaminated, it is difficult to duplicate aboriginal medicines. In the laboratory, research chemists and biochemists dare not use water directly out of the tap for their reactions. The reactions will fail. In the scientific community, we have known this for a very long time. The water used in a laboratory is double or triple distilled in glass distillation units. Lakes, streams, rivers, and, indeed, oceans are dumping grounds for chemical detritus, from raw sewage, to heart medications, to radionuclear chemicals.

Corylus, hazelnuts, were more common in America in the past.

Each living creature on this planet has inherited its own particular individual musical instrument. The music of life is played on DNA. Each species has its tune with its own harmonics: DNA is the musical instrument of life. We all take part in the symphony of life; our music is complementary. It is interconnected by the harmonics of sound itself. Our planet alone rings the bell of life. Our unique planet with its solid core produces oscillations or harmonics as it travels through the cathedral of our galaxy. It does this in an ever expanding swing, pushed by a force of outward movement that nobody understands.

I talk about *bioplans* and the act of *bioplanning*. These are words I coined for my last book, *Bioplanning a North Temperate Garden*, with a foreword by the Honourable Miriam Rothschild. In this book I said, "To bioplan a garden, one must realign the dimensions of the garden to encourage its use as a natural habitat."

In a garden this realignment can range from an act as simple as putting pinewood lollipop floaters in a birdbath so that the beneficial insects living in the garden can come and quench their thirst during a hot summer's day, to adding native plant species in the garden itself. In this time of global warming, shade should be provided in any garden and damp areas for our amphibian populations, such as frogs and toads. All these mindful acts are bioplanning; they extend the habitat of the garden to other creatures so that they may live and thrive in the balance that is nature.

In my garden book I define a bioplan:

> The Bioplan is a blueprint for all connectivity of life in nature. It is the fragile web which keeps each creature in balance with its neighbour. It is predation and prey. It is the victor and victim in a vast cycle of elemental life which is almost beyond our comprehension. It is the quantum mechanic of the green chloroplast without which we would all die. It is the domatal hairs on the underside of deciduous trees harbouring the parasites for aphids. It is the ultraviolet traffic light signaling system in flowers for the insect world. It is the terpene aerosol S.O.S. produced by plants in response to invasive damage. It is the toxin trick offered by plants for the protection of butterflies. It is the mantle of man, in his life and in his death, a divine contract, to all who share this planet.

To understand why we must enact the bioplan we should step backward a century or so to the beginnings of our technological era. Nature has been pried apart bit by bit. Concrete has replaced the lush rich soil. Gardens with a handful of petunias are supposed to feed migratory songbirds and butterflies with their need for high-energy nectars and seeds. The great forests and the aboriginal forest parklands of the North American continent are replaced by second- and third-growth brush in a patchwork of degradation where the forest animals are expected to cross busy highways in their search for family food and become the ubiquitous North American joke called "roadkill." Now, to survive as a species ourselves, we must put nature back together, we must hold hands, one with the other, and enact the bioplan. And like the stars of the heavens, each one of us must light our own pathway. It is only then that our combinations of connections will make the magic of the Milky Way. This we can do, not solely because we are just human, but because we hold the noblesse oblige of another's hand.

In these pages of *Arboretum America* I talk about the bioplan again. The forests must be repaired; they can be repaired by bioplanning. We are at the point where what remains of our forests are simply sterile arboreta for clinical examination. The vast swags of lichens hanging like sails from forest giants catching morning mists are almost lost even to our imagination. The array of diversity on the quaking mattress of forest floors and the primeval silence caressed by the fingering of rare fragrances are nearly gone. They must be brought back.

To bioplan a forest, quality trees of good genetic stock from the oldest, healthiest mother trees should be obtained locally. These trees should be many centuries old if at all possible. Their seeds are the progeny to start a forest from epicenters of quality, much as the squirrel has done for millennia. Then other quality companion trees are added, until a forest is born. Forests take faith, time and vision, in other words, a tincture of all the good qualities of mankind.

Design is fascinating to me. I have always loved art in all its forms. Indeed, I believe that art is the twin sister of science. Trees add to the art form of a garden. Like mountains and the sea, they change contours with the seasons. They enrich the eye, which feeds the mind and settles the soul to perceive the joy in a garden's change. I have suggested a list of interesting cultivars with this palette in mind.

This book was also written because I was sold a dream when I was six years old. The lessons of that dream have stayed with me all of my life. My mother had a favorite uncle, called Denny. He was a *cúipinéir,* that is a bone-setter, and a judge in the Gaelic-speaking world of southern Ireland. He was also a child of the great Famine and held the oral history of this, the greatest of Irish calamities, on the tip of his tongue. He lived by the Brehon Laws of Irish jurisprudence. His vision of the world, filtered by the smoke of plug tobacco, was that of sustainability. Every strawberry leaf was a source of a kidney remedy, every pennyroyal was for skin disease, every willow was for the making of sally baskets for hens laying in comfort.

One day, when I was six years old, he took me by the hand and we walked uphill to the giant Ogham stone above the farm. The smell of tobacco trailed after us. Upon reaching the stone, he swung me on top of it. I had a view of Bantry Bay and the Caha Mountains framing the blue waters and sky, making the green fields glow like small, square emeralds in front of me. Below us in another farm a field was being worked up. It bordered on a stream that fed the Ovan River, which tumbled, salmon-laden, out into the Atlantic Ocean.

"Child," he said to me, "mark this day." He peered down into the freshly exposed soils of the one brown square in an otherwise green quilt. "They're using superphosphate." He tapped his pipe, to empty it, against the ancient stone, until it sounded hollow. "You are looking at the rape of nature." He found a twig at his feet to clean his pipe. "The rape of nature," he repeated. He cleaned the stem. "So, child," he concluded, "mark this day." He blew out the stem until it was free. He swung me off the stone and told me again, "And . . . mark this day!" We walked home hand in hand, the brown field to our back.

Now, as a scientist, I have my own dream, that a moratorium be put on the cutting of what is left of the global forests and that ordinary people with an acorn and a shovel begin the long road back to nature.

Nomenclature

The Latin name of all plant species, as well as the common name, is used in the text. Where specific cultivars are mentioned, the cultivar name follows the Latin species name. For example, in *Acer saccharum* 'Laciniatum', the weeping form of the common sugar maple, *Acer saccharum* in italics denotes the species and 'Laciniatum' in single quotation marks, the cultivar. The genus, that is, *Acer,* always appears with the first letter in upper case, and subsequent species descriptions appear in lower case.

Generally *Hortus Third* was used for naming, definition, and spelling of plant genus and species names. In the case of the oaks, *Quercus* species, *The Hillier Manual* of *Trees and Shrubs* was used, as Sir Harold Hillier has a more complete collection of these species in the Hillier Arboretum. In addition, he is considered to be a world expert on this genus.

For each genus monograph, the following convention was used: first appears the genus name, followed by the common name, followed by the family name, terminating with the USDA climate zones in which the particular species of the genus, chosen for discussion in the text, can be expected to grow.

An example is as follows:

Genus	*Acer*	
Common name	MAPLE	
Family	Aceraceae	Zones 3–10

THE TREES

Caution

To all readers of *Arboretum America: A Philosophy of the Forest:*

The aboriginal medicines described in this book cannot be duplicated exactly. The chemical reason for this is the worldwide use of toxic pesticides. It has just recently been discovered that many pesticides are carried either in a mechanical or molecular aerosol form and are distributed in a pattern dependent on the shape, size, and density of the pesticide itself. This contamination is worldwide in the global garden; no region escapes, from the Himalayas to the poles. The synergy of pesticides with both themselves and the medicines of the plant kingdom is to date unexplored, unknown, and therefore potentially dangerous outside of an experimental laboratory.

Acer

MAPLE

Aceraceae

Zones 3–10

THE GLOBAL GARDEN

The most ancient ritual of North America is the production of maple syrup. This unique tree sugar is the survival trick of one family of trees placed on a razor's edge of hunger, a hunger primed by scorching sun and blistering winters of the north. The family name is maple, Aceraceae.

The prince of the litter is the sugar maple, *Acer saccharum*. Outside of the biblical manna, nothing has ever been before or will ever be again so gifted in food production as this common, native tree.

The culture that followed the custom of sugaring has been part and parcel of North America for a very long time. Knowledge of the maple was by no means unique to this continent. The custom probably trickled over the Bering Straits twenty thousand years ago. Maybe it came from the ancient kingdoms of China, for their people certainly knew how to tap "sweet water" from their maple species. Maybe, somehow it came westwards, later, with druidic knowledge, because maple is called *Mailp* in ancient Gaelic and is a very old word. Even the Druids knew how to tap the maple. But it was the First Nations who hung their customs and rituals on this tree. Each spring the Mohawks venerate the maple as the leader of all trees. In their language they call it *Wahtha*.

The richest traditional relationship with the sugar maple

Autumnal incandescence. It has never been fully explored by horticulturists and gardeners in North America.

Flowers of *Dicentra canadensis,* squirrel corn. Their fragrance calls to the early bumblebees, whose proboscises are just long enough to reach the rich nectar. The roots, resembling corn kernels, hug the trees for additional warmth to grow. This species is an aboriginal medicine for long-distance runners.

belongs to the Chippewa peoples. Unfortunately, these rituals are getting lost in time. Their average extended aboriginal family needed about nine hundred taps to meet their yearly supply of maple sugar and its products. Each maple, depending on its girth, had a given number of taps. The aboriginal Chippewa sugaring operations began, as they do today, around the middle of March and ended about a month later. Their first run, or *grand cru,* was produced when the first part of the winter was an open one. This was always when the frost went deeply into the ground and, following on its heels, a lasting blanket of insulating snow covered the soil.

Their full operation of collection, boiling, and condensing was done using utensils made out of birch bark. The birch and the cedar are the two sacred trees given to the Chippewa by their legendary heroic figure Winabojo. The birch lives up to this status, because the inside of the bark, that is, the green side next to the xylem-wood, is fireproof. Sheets of birch bark were slipped in the spring. They were then heated to make them pliable, after which they were fashioned into waterproof and flameproof kettles. This bark is also antimicrobial or antibiotic, and so long-term storage is clean and sterile for the sugars stored in the packages called *makuks,* also fashioned from birch bark in differing sizes.

The *makuk* is almost identical to a briefcase. The syrup was reduced over an open fire using the birch kettles. Then a little dollop of deer tallow was added to keep the maple sugar flexible but not crumbly. For long-term storage, this sugar product was packaged into large *makuks* kept in mobile storage huts.

Maple syrup was an item on the daily menu of the First Nations throughout the millennia. It was used to cook fish, meat, and game. It seasoned their vegetables such as *Helianthus tuberosus,* the Jerusalem artichoke; *Sagittaria latifolia,* the arrowhead (an aquatic tuber); and the fresh acorns of *Quercus macrocarpa,* the bur oak. These nuts were split open and used as a vegetable. Maple syrup was also used to sweeten fruits in season. These included *Crataegus mollis,* haws; *Prunus virginiana,* chokecherries; *Cornus canadensis,* bunchberries; as well as their all-time favorite, the little wild strawberry, *Fragaria virginiana.* The choice, dark cereal of the north temperate lakes and backwaters, wild rice, *Zizania aquatica,* was boiled using maple sap much as white rice is cooked today in water.

The first flux of settlers rapidly discovered that the American maples had much in common with the dwindling stocks of English maple, *Acer campestris,* they had left behind. They knew charcoal and had cooked with it. It was common knowledge that the ash content of the maple is one of the highest of any wood, being about 4 percent of the dry weight. This was dubbed the "Black Gold" of North America, spelling a pioneer fortune of $15 per full cord, or $4.14 per cubic meter. The race was on to fell the vast stands of maple to produce charcoal for export back home, to feed the glass, gunpowder, and rag industries of Queen Victoria's England. When charcoal became illegal to sell in the United States of America, it was smuggled north into Canada and then into England. The ash houses of Canada and the impressively huge ash houses of landed estates in England are the historical reminders of this trade today.

For the motley pioneers who had to suffer "the stings and arrows" of outraged bees for a little honey to sweeten tea, maple syrup was love at first sight. So the odd maple was saved, proving that the sweet tooth is mightier than the ax. Farms were described by the "number of taps" as well as the acreage. Nowadays, moss-covered stone fire pits, looking like long graves, are still to be stumbled over in Canadian forests from bygone, crude maple-syruping operations. And the shoulder yoke for syrup carrying is now replaced by a high-tech vinyl tubing as syrup is gathered effortlessly by gravity and pumped into a multi-million-dollar, virtually unique, Canadian business providing some 75 percent of the world's maple sap products.

As a family, the maples are the giants and the pipsqueaks of the forest, ranging from the 100 foot (30 m) giant sugar maples, *A. saccharum,* and the black maples, *A. nigrum,* to the slimline shrubs called moosewood, *A. pensylvanicum,* and the gnarled little vine maple, *A. circinatum.* This tiny tot is related to the Japanese maple, *A. palmatum.* Another maple caught in the net of evolutionary flotsam is the big leaf maple, *A. macrophyllum,* of the North American west coast rain forest. This tree, for some inexplicable reason, has family links to the one and only English maple, *A. campestris,* now a rarity of pleached hedges in the English countryside. Also reaching into the prairies is the roughest of them all, *A. negundo.* This tree is called the Manitoba maple in Canada and the box elder in the United States. Stepping into these harsher regions as well is the red maple, *A. rubrum,* and the understory shrubby mountain maple, *A. spicatum.*

ORGANIC CARE

All maples can be grown from seed. There are one hundred or so species from which to choose. They all have a similar pattern of growth. The seed is produced as a twin, in form, like the blades of a helicopter. This twin is a fruit called a *samara.* The samara can be snapped in two to produce two seeds, each with a blade or wing. These seeds can be expected to grow into a maple tree.

The timing of the collection of maple seeds is important for the grower. Unlike most trees of the north temperate forests, the maple produces seeds precociously. The trees flower with the first hot breath of spring, and the fertilized fruit is ready to drop in early summer. The seed is then to be collected, when the tree is in full leaf.

Trillium grandiflorum, large-flowered trillium. It helps to form a white, spring apron in the maple woods. An oil of trillium was used by the aboriginal peoples to treat rheumatism.

The seed crop of the maple varies greatly from year to year. This depends in part on the tree's very strange sexuality and on the eccentricities of a variable, continental climate. The sexual web thrown out by the maple includes everything in the book, heterosexuality, homosexuality, and polygamodioecious behavior where the pollen load is formed and released before the female flowers are ready to be courted. This singular pattern produces, as would be expected, few offspring. Sexual deviance is just part of a genetic master plan for stalling, keeping the tree inside the fold for survival.

The samaras for the moisture-loving maples, *A. rubrum,* the red maple, and *A. saccharinum,* the silver maple, are collected as soon as they are plump and green in midsummer. These samaras will grow as soon as they have been collected.

All other maples are collected and allowed to dry down in a sunny place to a color that is reddish brown. Open wickerwork baskets are ideal for this. The samaras, as they dry, become

Speak to Me

Pushing my baby daughter down the leafy arcade of our driveway, I was racing away from the sound of crying, howling, and teething that with which all mothers are so familiar. I stopped. I left her under a huge maple tree. I put my hands to my face. I wanted to cry, too. I began to walk backward up the road. Suddenly a silence came into my head. It was followed by a rushing train of unspoken words in a silent speech that was coming from the huge maple tree. I ran to my child. I grabbed the handle of the stroller and furiously dragged it backward with all my might. She had stopped crying. As I managed to pull her clear, I heard a huge thunder overhead. The main limb broke. It would have killed her, that is absolutely certain. Who gave me that warning? The tree or me. I just don't know . . .

lighter and are able to fly. The color change indicates that they have reached a moisture content of about 15 percent, which is ideal for storage. They are collected into covered glass jars and are stored in a refrigerator at 41°F (5°C) until spring planting. Sugar maple, *A. saccharum,* stored in this manner, will be viable for up to five years.

The west coast rain forest maple, *A. macrophyllum,* is collected in midsummer and stored at 41°F (5°C) for just forty days. It is then ready to plant outside the same fall.

The maple embryo is kept dormant or asleep by moisture, which in turn affects the size of the embryo. When the embryo is fat and bulging, it cannot grow, but if it is left to dry down to a smaller size, the embryo senses space around it. This is the trigger for growth.

All maples like a rich, well-drained soil. Most of the maples prefer a slightly acidic soil with a pH 6.0–4.5. The sugar maples, *A. saccharum,* and the black maples, *A. nigrum,* being the exceptions, like a soil pH 6.5–7.5. Both of these trees thrive on soils with an underlying formation of dolomitic limestone.

The maple samaras are planted flat with each individual wing intact, ¼–1 inch (.6–2.5 cm) deep. This is done in the fall. In addition a light dusting of dry woodash is powdered over the nursery row to inhibit the growth of pathogenic fungal spores.

All maple seedlings are easily transplanted. In their first two years of growth some shade helps to make these trees vigorous. This can be provided by one layer of spun polyester row cover such as Reemay (a trademarked name). It can also be supplied by a nurse crop of annuals like yellow sweet clover, *Melilotus officinalis,* or by other trees.

In thin, poor soils maples are prone to disease, especially a fungus called verticillium wilt caused by *Verticillium dahliae* and *V. albo-atrum.* Many strains of these fungi exist. They are becoming more virulent because of global warming. With an increase of temperature and carbon dioxide, global soils are becoming more acidic, favoring the wilt. To protect all maples, especially the large species, from this effect, ground dolomitic limestone can be spread around maples at the rate of 2 qts/100 ft^2 (2 l/10 m^2) per tree. This can be hand broadcast in a gridlike fashion out to the drip line of the tree.

This wilt can also find entry through careless pruning of maples. All maples are pruned in the early winter only, when they are fully dormant. This is December in the east and early January on the west coasts of North America and Europe. Any sap leakage will invite pathogenic fungi to enter the tree. For the maple, a ¼–½-inch (.6–1.2 cm) heel is left on the branches or trunk to regenerate callus tissue. Both the cut and cutting tools are sterilized with either alcohol or a 2 percent solution of water and household chlorine bleach (50 parts of water to 1 part of bleach). This kills invading spores.

MEDICINE

The medicine of the maple is found in the bark, the leaves, the twigs, and the sap, particularly the sap of the sugar maple, *A. saccharum.* The sap of the sugar maple contains eight major biochemicals that are always in a constant proportion and never vary despite the geographical location of the trees. These biochemicals are the identifying tracers for true maple syrup.

The first flow of the fresh, raw sap in the spring has long been considered to be a general physic or tonic in North America. The aboriginal peoples drank about 2 pints (.9 l) for this purpose. This medicinal practice is still in use at the present time, particularly in Quebec.

The sap of the sugar maple, *A. saccharum,* has acetol, isomaltol, clyclotene, α-furanone, hydroxymethylfurtural, vanillin, syringaldehyde, and dihydroconiferol alcohol. This sap has a diuretic effect with a cleansing action on the skin. It has a positive effect on the spleen. Acetol acts as an antistroke agent by having a hemodilution effect on the blood. Vanillin and syringaldehyde impart taste and fragrance to the palate. The compound α-furanone has an antibiotic action and decreases cholesterol levels. The synergy of the combined chemicals in the sap no doubt have an additional beneficial effect on the body. It is the reason why this folkloric medicine is still being used today.

In addition, the Manitoba maple, *A. negundo,* appears to have a biochemical that is active against cancer. This anticarcinogen appears to be a complex saponin that is at present under investigation.

The aboriginal peoples of North America used the red maple, *A. rubrum,* in treating ailments of the eye. In particular, the treatment was used to delay or retard the formation of cataracts. A young sapling of red maple was chosen in the spring. This was when the young leaves had turned from green into the beginnings of the red phase. The bark diameter was about 2 inches (5 cm). This was scraped upward, starting from the midpoint of the sapling. A small handful of bark was taken. This was steeped in 6 imperial fluid ounces (170 ml) of warm well water. This water had been previously boiled and allowed to cool down to lukewarm or blood heat. Two drops of this solution was carefully placed into the eye.

The First Nations made much use of the pharmacy of the moosewood tree, *A. pensylvanicum.* It was used as a laxative. A handful of bark was scraped upward. This bark was covered with well water and brought to a boil. The resulting decoction was cooled and drunk in quantity before breakfast.

The Iroquois aboriginal peoples used the red maple, *A. rubrum,* in a fascinating way—as a deodorant. Two to three handfuls of young shoots were placed into boiling water and then removed. The water was cooled a little and then their trapping implements were dropped into it. This process deodorized the implements of all human scent. They called the red maple a trapping medicine.

Scientists in Europe have extracted a biochemical called acerin from the dried fruits of the Norway maple, *A. platanoides.* It is being investigated for its antiviral and antibiotic status. The acerin appears to be quite similar to ordinary vegetable tannins, but the molecule seems to have an unusual plasticity. Similar chemical structure might be found in the rocky mountain maple, *A. glabrum* Torr. var. *glabrum,* or the vine maple, *A. circinatum* 'Monroe', of the Willamette National Forest, Oregon.

Allium tricoccum, wild leek or ramp. Unknown in Europe, it is a medicinal companion to North American maple woods. It has been used since ancient times as a spring tonic.

ECOFUNCTION

Some species of maples have killer instincts. Their strategies would win Agatha Christie awards. These plotting species are the black maple, *A. nigrum,* the sugar maple, *A. saccharum,* the bigleaf maple, *A. macrophyllum,* the Florida maple, *A. barbatum,* and the English maple, *A. campestris.*

An *Acer saccharinum,* silver maple, that gets its size from an artesian well under the house. The house dates the tree by being well over a hundred years old.

When a large living tree is cut down, usually a wedge is cut out of the bole and the tree is felled. However, ticking away in the killer maples is a fungus called *Cryptostroma corticale.* It lives between the epidermis and the endodermis of the bark, residing primarily in the cortical tissue, which is a housing complex for phloem cells and food storage next door. This cortex could be compared to the thymus in man, active in youth and going with age. The *Cryptostroma* fungus acts in synchrony with the microscopic hair fields on the mature leaves. When the maple falls, these glandular hair tips shatter like glass, flying through the air and, like asbestos fibers, become part of the breathing mixture of the logger. The tips of the hairs are extraordinarily fine. They puncture the alveolar tissue of the lungs, carrying the minute spores of the fungus, *Cryptostroma corticale,* as a killer, aerodynamic load. The result is a disease called extrinsic allergic alveolitis, which can be a deadly allergic respiratory condition. The only living creature of the forest that seems to have some kind of cunning about this potential killer is the lowly porcupine, who confines his maple snack to the spring sans hairs.

The red maple, *A. rubrum,* is not so benign either. This tree produces powerful antifeeding compounds that can kill both cattle and horses. Here again the deer seem to have some smarts; they only winter browse on this tree, a condition that matches the winter switch of their intestinal flora.

A growing maple, anywhere, is a source of uncontaminated water. In the spring the squirrels uncork the buds, eating them and feeding on the sap. The winter birds, such as the black-capped chickadees, *Parus atricapillus,* move in for a drink, as do the early butterflies. As the air warms up, the ants get wind of the sweetness and stick around. Close on their heels are the woodpeckers. The *Dendrocarpas* species drill out the ants, dine on them, and in the act create their own homes. These holes get enlarged, and forest mammals, like the flying squirrels, *Glaucomys sabrinus,* move in communally for the winters. The flickers, *Colapets auratus luteus,* lurk around to move in and raise their large broods. They feed on ants until a portion of the tree hits the ground.

Fitting into this northern symphony are the porcupines, *Erethizon dorsatum*, who feed on the buds, twigs, and fresh bark of late winter to early spring. Overwintering porcupines like to take up communal winter dwellings in the hollowed trunks of ancient maples. On the warmer days they enjoy taking an occasional jaunt up the branches to do a little snacking. Their volatile, urinal plume can ungracefully identify their maple home in the forest. But, this, too, compensates the tree with an intensive fertilizer, paying for the porcupine's sin of sweet, cambial gluttony.

In the early spring when the canopy is absent, the sun's heat reflected among the maple trunks produces a wildflower woodland of magnificent beauty. These woodland plants are cormous, bulbous, or rhizomal, all maintained in health by the calcium recycling of the maple tree litter.

Dominant are the trilliums, which are the ancient virgin forest species. These are followed by the ferns and the fruiting bodies of many varieties of mushrooms like the *Polyporous* and the white *Lentinellus* species that move onto the fallen limbs and add a considerable tilth to the forest floor.

There is a generous pollen production and nectar flow from the spring-flowering maple tassels. Depending on the location and weather, insects can make ample use of this early food source.

BIOPLAN

The larger maples should not be bioplanned around hospitals, clinics, retirement homes, schools, or day-care and nursery centers. This should also include suburban areas of small towns and cities that are downwind of nuclear reactors. In these areas female trees are to be preferred as shade trees because they do not produce pollen. In addition, any of the smaller maples can also be planted. Because of their reduced size, they produce negligible amounts of spring pollen.

The male, or pollen-producing maples, especially in built-up centers with a high proportion of hard surface areas, accumulate wind-blown pollen in moving, surface shoals. These shoals cause hay fever or allergic rhinitis for much of the population. The pollen is inhaled into the damp mucosal membranes of breathing and becomes an allergen. This situation is exacerbated in built-up areas and downwind of nuclear reactors, which occasionally gas off small quantities of radionuclear material like iodine that can affect the thyroid in its functioning. If the person is further exposed to allergens, the thyroid is increasingly depressed, especially in those who are predisposed to thyroid dysfunction.

Maple pollen shoals are harmless in the countryside because the pollen grains, being hydrophilic, are collected on the damp uneven soil surfaces. Both dew and rain wash the pollen into the soil, where it is quickly consumed by soil organisms.

At present the dwindling maple population is not being replaced. Genetically inferior imported tree stock that is easier to manage in the sapling stage is being used as a substitute. There should be a "set-aside program" for the farming community to plant these trees as part of a tax incentive to encourage the planting of native species. The efforts of the Irish government could serve as a model in this regard.

The maple is connected to a cycle of carbon dioxide reduction globally. In the north temperate forests, the large maples and their deciduous colleagues switch on photosynthesis after winter dormancy. This cycle can be observed as a wave pattern of atmospheric carbon reduction during the summer and early fall months. With global warming and the loss of the forests this cycle of carbon respiration is beginning to plateau. This is one living key to the cause of the greenhouse effect. It can be reversed by the informed actions of ordinary citizens.

DESIGN

Horticulturists and garden designers have never fully exploited the autumnal incandescence in North America. These color changes are more vivid and striking in the various native maple species than anywhere else in the world. Their brilliance is the hallmark of a north temperate garden.

Many species are available for woodland gardens. These species also work for larger shade gardens and country estates. The most famous is the sugar maple, *A. saccharum.* This native turns bright red to gold. A more showy cultivar of this maple is *A. s.* 'Globosum', which offers brilliant red tinges on the golding leaves. The canopies of these trees can be trained to meet and form a luminescent walk that lasts for two or more spectacular weeks. By way of a large tree, the red maple, *A. rubrum,* turns from blood red to gold. Its cultivar, *A. r.* 'Schlesingeri', slips into autumnal color two weeks earlier than other maples but holds its color longer. In recent years the Manitoba maple, *A. negundo* 'Variegatum', which has a variegated leaf form, is being used as a specimen tree. However, after the summer's peak of growth the foliage goes into a premature brown ebb. A maple look-alike that is a sycamore cultivar is being increasingly used in southern gardens with great success, especially from midsummer onward. It is *A. pseudoplatanus* 'Brilliantissimum'. The foliage turns pink, yellow, green, turning into autumn gold.

Small gardens can lay claim to fantastic fall color from the hardy, stipple-trunked moosewood maple, *A. pensylvanicum.* This too, has a truly, but seldom used, spectacular cultivar, *A. p.* 'Erythrocladum', with yearlong, bright red twigs, which, when loaded with swags of samaras, make a colorful spot in a winter garden. For the tiny garden there is the Chinese maple called the paperbark maple, *A. grisum.* The outer bark of this tree peels back in bleached curls to reveal a rich, cinnamon, velvet underbark.

The popular Norway maple *A. platanoides* 'Crimson King' is marginally hardy in zone 4. But the miniature Japanese maples, *A. palmatum,* and their dwarf tribe of cultivars are far more hardy than is generally realized. These will grow in zone 4. Few trees grow well in damp or wet soils better than the silver maple, *A. saccharinum.* There is a magnificent, cut-leaf cultivar, *A. s.* 'Laciniatum', and an upright, noble *A. s.* 'Fastigiatum' cultivar of the silver maple that is seldom seen but much admired when grown.

Simplicity is the key for the maple. A wildflower apron of pure white in the spring will be followed by the color flow of the foliage in the fall. With the first warmth of spring, meadows of Dutchman's breeches, *Dicentra cucullaria,* with delicate fernlike foliage and nodding rows of flowers interspersed with the later blooming, giant wake robin, *Trillium grandiflorum,* state the sparse architectural majesty of the maple tree from its moss-packed base to its powdery red tassels in a manner that can only be of a north temperate garden.

Acer Species and Cultivars of Merit

SCIENTIFIC NAME	COMMON NAME	ZONES
Acer circinatum	Little vine maple	4–9
A. grisum	Paperback maple	4–9
A. macrophyllum	Big leaf maple	6–10
A. negundo	Manitoba maple	2–9
A. n. 'Variegatum'	Variegated Manitoba maple	2–9
A. nigrum	Black maple	4–9
A. palmatum	Japanese maple	4–9
A. pensylvanicum	Moosewood maple	3–9
A. p. 'Erythrocladum'	Red moosewood maple	2–9
A. platanoides	Norway maple	5–9
A. p. 'Crimson King'	Crimson Norway maple	5–9
A. rubrum	Red maple	2–9
A. r. 'Schlesingeri'	Schlesinger red maple	2–9
A. saccharinum	Silver maple	3–9
A. s. 'Laciniatum'	Cutleaf silver maple	3–9
A. s. 'Fastigiatum'	Upright silver maple	3–9
A. saccharum	Sugar maple	3–9
A. s. 'Globosum'	Globose sugar maple	3–9
A. spicatum	Mountain maple	2–9

Lobaria pulmonaria, lungwort. Once upon a time in the ancient virgin forests of North America, giant sails of lichens swung from the trees to trap rain and moisture.

Asarum canadense, wild ginger, a humble companion to *Asimina triloba,* pawpaw, is a tonic and a physic. It was used to regulate fevers in North America. *Asarum* species are medicinal in all of the global garden.

Asimina triloba

PAWPAW

Annonaceae Zones 4–10

Asimina triloba, pawpaw, which takes its stride in zone 4. It was hardiness tested with its sisters and brothers. This tree wins the gold medal as the toughest of the tribe.

THE GLOBAL GARDEN

The poor man's banana grows in Canada. It is a fruit that has a flavor of a mixture of mango, pineapple, and banana. The fruit occurs in bunches with weights tipping the scales at 2 pounds (.9 kg). In fact, it is the sleeping giant of the North American forests, quietly snoozing its life away in the cool of the shade, a snoring treasure trove that goes unnoticed by scientists, government, and the general public. It is said that some international pharmaceutical companies are very aware of its potential, but the time is not ripe to open the chest. Long ago the giant was prodded by the aristocratic Hernando de Soto, a gentleman with an inbred taste for the exotic. The delicate fruit passed his lips in the Mississippi valley. But that was in 1541, far too early to be noticed in the noise and havoc of the twenty-first century. In any case the local aboriginal peoples at that time knew about the fruit but were inclined toward silence because it was also their medicine. History, it turns out, has a permanent means of silencing the vanquished. The raccoons and opossums know about the fruit. They would have spilled the beans quite happily. But they, too, are silent, being no friends of man!

The "giant," or the tree in question, is only a small one. It is commonly called the pawpaw, the custard-apple, or the common pawpaw. It is a member of the Annonaceae family and is called *Asimina triloba.* The largest tree on record is about 55 feet (17 m) tall with 27 feet (8 m) of spread. In reality the pawpaw is a temperate member of a tropical family to which the common sumac, *Rhus typhina,* is related. This family includes several exotic fruits that are familiar to globe-trotters. These are the custard apple, *Annona reticulata;* the sugar apple, *A. squamosa;* the soursop, *A. muricata;* and the cherimoya, *A. cherimola.* All of these species are native to the extreme southeastern United States from Georgia to Florida. However, the pawpaw, *Asimina triloba,* is found growing naturally in southern Ontario and western New York state down into the Gulf of Mexico and westwards into Nebraska. It also grows quite happily into elevations of 3,000 feet (900 m) in the southern Appalachian Mountains. This tree appears to have been stranded in its present habitat by the last ice age. The tree has, since then, modified its genetics naturally to enable it to thrive in a colder climate. This is probably the reason why the pawpaw grows in England, both setting fruit and ripening in that climate. The pawpaw, as one would expect, grows well in Chile and also in China.

The pawpaw, *Asimina triloba,* is a Pandora's box inside which are astonishing new medicines, a range of natural pesticides, a new delicious fruit, a new flavoring, a new cosmetic aroma, a new greenhouse horticultural species, a cash crop for the farming community, a bonanza for the scientific community, a boon for many major industries, and a definite challenge to organic chemists specializing in stereochemistry.

ORGANIC CARE

Asimina triloba, the pawpaw, is native only to North America. This tree can be grown quite readily from seed. The seed is large, about the size of a broad bean, and is easily handled. The pawpaw can be propagated by whip, whip and tongue, bark inlay, and chip budding. These methods are currently in use in the apple business. Presently, research is being conducted on hardwood, softwood, and root tissue cloning. When the pawpaw is being grafted, only the largest vegetative buds should be selected. Flower buds should be avoided. The diameters of the scion pawpaw wood and the receiving pawpaw mother wood should match. This results in a good "take." Micropropagation techniques have also been successful, the best source of tissue being fully dormant twigs. Nodal cuttings from the twigs have been found to be a superior source of explants.

Pawpaws can still be found growing plentifully in the wild in a few areas. They are usually found in the shade as an understory in damp places, such as by running streams. The tree has a tendency to clone in its native habitat and resents being disturbed in any way. This is because the roots are brittle and when broken do not seem to regenerate very rapidly. These remaining native species should be respected and left alone. However, they can become a seed source if needed.

The gardener, farmer, or greenhouse operator can easily and cheaply grow the pawpaw from seed. These seeds are collected as soon as the fruit is soft in the fall. The fruit, if left to ripen, will ferment. This seems to damage the immature embryo inside the fruit and will prevent its germination. The collected seeds should not be allowed to dry out. They can be stored in plastic. However, the dry seed coat is prone to fungal infection and a rub with a light-grade vegetable oil will seal the coat and considerably increase germination after storage.

Asimina triloba, the pawpaw, requires a minimum of 100 days of stratification of its seeds. This process further matures the embryo, making it ready for the trials of germination. To do this, the seeds are stored at 41°F (5°C) in a household refrigerator. After this time the seeds are planted ½ inch (12 mm) deep in soil. The soil should be amended with one-quarter volume of sand to aid in the seeds' hypogeal germination. The hilum on the seed is identified. It is the slightly rough, triangular point of attachment to the ovarian placenta of the fruit. This hilum is placed downward. This will increase the germination of the fruit. The potted seeds are plunged into sand and covered with a layer of spun polyester row cover (Reemay) during the first winter. The majority of seeds will germinate when the soil temperatures increase in the spring. For the warmer zones of 7–9, this will be any time from March to April. For the colder zones of 4–6, this happens in May. Pawpaw seedlings have a remarkably uniform timing of growth and robust character.

Young pawpaw seedlings require two items for continued growth, shade and a plentiful supply of water. Reemay used over the seedlings both shades and protects them from spurious nighttime frosts. Direct sunlight seems to have a damaging effect on the juvenile growth of the young seedling. In the warmer zones with higher solar exposure this is more of a problem. As the seedling grows, it shows a greater accommodation to the sun. The leaf design of the pawpaw is such that it transpires, or loses water, at a rapid rate, so water must be in good supply.

The pawpaw seedlings grow rapidly into a sturdy, 6-inch (15 cm) tree in about twenty days. As the summer progresses, it grows a little stout before hardening off naturally in October or early November. At this time, the young trees can readily be transplanted out for zones 7–10. For zones 4–6 a spring transplanting in April to May is preferred.

Pawpaws like and bear heavily in a rich, damp, warm soil with a pH 4.5–7.0. With these soil conditions the tree will grow in shade or in the open. The soil should have good winter drainage. To increase bearing, when transplanting, an adequate amount (1 qt, .9 l)—of steamed bonemeal should be added to the planting hole. Dolomitic lime—(.5 qt, .5 l)—should be added also. Both fertilizers should be mixed well into the soil before it is placed below the tree roots. Woodash (.5 qt, .5 l)—can be sprinkled on the soil surface to the drip line after the tree has reached three years of age. This will increase the fruit quality and size, especially in a northern plantation, and will also increase the winter frost protection of the plant itself.

Pawpaw seeds can be stored in plastic at 41°F (5°C) in damp peat moss for five years and longer. They will retain their ability to germinate for at least this period and possibly much longer.

MEDICINE

There has been a tradition in the southeastern areas of the United States of going for a fall pawpaw hunt. The hunt was carried out by those who lived close to the land in places like the Appalachian Mountains. These small farmers and settlers also had another name for the pawpaw. They called it the frost banana. The settlers had learned of the value of the hunt from the Cherokee and Seminole peoples into which many had intermarried. The fruits were collected when they had turned from yellow to brown. They were eaten like a pear. They were also baked, put into pawpaw pie, made into pawpaw flump, and the pulp was added to nut breads. This gave the cooked bread an attractive rose-red color. The reason for this fall hunt was that the pawpaw is a superior food. It exceeds the apple, the peach, and grapes in most vitamins, minerals, amino acids, and energy value.

Few people now remember the pawpaw hunt of an earlier age and even fewer remember the medicines attached to a tree that has almost disappeared due to habitat destruction. No less was

Asimina triloba, pawpaw, taking on the characteristics of a shrub in Canada

Oenothera odorata, evening primrose. With its long, yellow, night-fragrant flowers aging to red, it is a true tropical companion to *Asimina triloba,* pawpaw. This, too, is a medicinal plant.

the destruction of the Cherokee peoples, whose vast pharmacological knowledge of this species went with them into their graves. The little that is known of the aboriginal uses of the plant is that the bark of the *Asimina triloba* could be threaded into an excellent line used for fishing. In addition, a beverage that was known as Seminole tea was made from the flowers of a very closely related species called *Annona reticulata.* This tea was used to relieve kidney troubles. It was a medicine of the Seminole peoples of Florida.

There are as yet two undistinguished botanical races of *Asimina triloba.* One race produces a fruit with a rich, yellow pulp with a high flavonoid content that yields the wonderful mango-pineapple-banana flavor. The other race produces a pale yellow-fleshed fruit that tastes like turpentine. Both races have uses in medicine.

Asimina triloba, the pawpaw, holds a new class of chemical compounds that were previously unknown to science. These are presently being called annonaceous acetogenins. There are over forty of them in *Asimina triloba* alone. The other related tropical *Annona* species are yielding more acetogenins with investigation. Acetogenins are important because they represent a new class of long-chain fatty acid derivatives. They have a chemical architecture that makes them unique. In the human cell, in particular, they can inhibit cell metabolism. They block an extremely important enzyme called nicotinamide adenine dinucleotide (NADH). These enzymes are in the mitochondria, which are the powerhouses of the cell. The other enzyme that is affected by the acetogenins is a widely found one in the cell. It is called nicotinamide adenine dinucleotide oxidase. This enzyme sits in the plasma membranes of sixty different types of human cancer cells. This is according to the in vitro studies to date.

The anticancer action is very simple. It is based on the deprivation of the cell's energy, which is called adenosine triphosphate (ATP). In other words, the cancerous cell is like a car with a full tank of gasoline. The annonaceous acetogenins drain the tank so that the car cannot run. The result is that the cancer cells die from a lack of energy.

To date 250 different annonaceous acetogenins have been discovered or chemically characterized. These compounds are important because they represent a new kind of medical weapon that is active against drug-resistant cancers. Presently they are also effective against malaria and microbial and other parasitic infections. These infections will be on the rise due to the mutation effect and temperature changes of global warming. Analytical, chemical, and medical research has almost come to a halt on *Asimina triloba* due to a lack of funding.

ECOFUNCTION

Asimina triloba, the pawpaw, is an excellent lesson in the importance of biodiversity. It is a warning to us to be careful. Because if you destroy the forest, you destroy the pawpaw understory. If you destroy this understory, you destroy the specialized beetles for pollination. If you destroy them, you destroy the fruit. If you destroy the fruit, you destroy the yellow-billed cuckoo that uses it for food. If you destroy this bird, you destroy the distribution of the seed. If you do this, you destroy the host plant for the zebra swallowtail. If you do all of these things, you destroy the molecular protection for that swallowtail and the molecular protection that the lowly pawpaw can spin into the chemistry of man against cancer.

The beanlike seeds of *Asimina triloba* are quite large. The tree uses small boys and tactile animals in an exquisite trick of dispersal for these seeds. When held in the hand (or raccoon's paw), the seed fits comfortably into the palm. The natural inclination is to turn it over and over because of its smoothness and bulging shape. It is the ideal "worry piece." The act of turning it over and over rubs skin lanolin into the seed coat's leaky testa and seals it against fungal infection. The oil sealing and the movement activates the internal, embryonic clock into germination for when the seed will be sewn in the ground. So begins its new cycle of life and species survival.

High-quality fall foods are essential for migratory birds in general. The richer the food, the greater is its reproductive potential. The yellow-billed cuckoo, *Coccyzus americanus americanus,* with its rufus wings and white spots on the tips of its tail feathers, loves the fruit of *Asimina triloba.* This bird breeds in New Brunswick, Quebec, Ontario, and down into the Florida Keys. It is not just pesticides that are killing this songbird; it is a serious decline of habitat and quality food that that habitat provides. Increased pawpaw plantings would improve its ability to survive its migrations into South America.

The *Asimina triloba,* the pawpaw, is the only host plant for the kite swallowtail, which is also called the zebra swallowtail, *Eurytides marcellus.* Like many conspicuous butterflies, they need avian protection. This, no doubt, is supplied by the chemical mix in the pawpaw's leaf.

The maroon flowers of the pawpaw are pollinated by beetles. The flower, when it is fertile, produces an attractant chemical that has a fetid odor. The beetles are attracted by this odor and the flesh color of the flower itself. This trick results in fertilization of the flowers and the production of viable seeds.

On leaf injury, or on the bruising of the twigs, the pawpaw produces an ill-smelling antifeeding compound that is the basis of its potential as a natural pesticide.

BIOPLAN

Asimina triloba, the pawpaw, represents a new cash crop for the crippled farming community. The tree can be readily grown in zones 4–10 with hardiness trials being conducted for zones 2–3.

Pawpaws can be coppiced. This is an old English technique for obtaining biomass indefinitely, for centuries if need be, without killing the tree. The stump or stool continues to live and the tree regenerates rapidly. The biomass could be harvested at peak growing times in July. It could be baled and transported for the manufacture of natural pesticides. This is presently being done in England with the English yew, *Taxus baccata.* Extracts of the *Asimina triloba* biomass are active against many pest species, including mosquito larvae, European corn borers, spider mites, melon and green peach aphids, Mexican bean beetles, bean leaf beetles, striped cucumber beetles, cabbage loopers, nematodes, blowfly larvae, Colorado potato beetles, sweet potato whitefly, and cockroaches.

The trees could also be grown in pawpaw orchards. The fruit could supply many industries such as the blended fruit drink industry and the ice cream, yogurt, and soya industries. Exotic fruits have always been a popular source of fragrance in the huge cosmetic industry for hair shampoos, soaps, bubble baths, and skin conditioners. Because pawpaw pulp freezes well and retains its flavor, it can be used in the fast-food industry as well as being an additional, exotic, North American fruit in the fresh fruit market. The fruit pulp both travels and stores well.

Some standards have also been set up for the greenhouse industry. Because *Asimina triloba* comes equipped with its own internal pesticide regime, it would be a cheap and easy switch for the maritime tobacco farmers to make since many greenhouses are lying vacant at present.

Asimina triloba should be used as an understory habitat

Fritillaria persica 'Adiyaman' repeats the dark tones of the pawpaw, *Asimina triloba,* flower.

Muscari comosum 'Plumosum', feather hyacinth, which hails from Western Europe. It can be used to great advantage with *Asimina triloba,* pawpaw, in a garden design. This edible bulb is sometimes sold in the United States as *Muscari cipollino* Cv. 'Monstrosum'.

restoration species from Canada down into the Deep South. Quite often it is found as an understory tree with mature *Cornus florida,* the flowering dogwood. When the dogwood has flowered, the pawpaw begins to unfurl its leaves.

The pawpaw should be grown in gardens, especially shaded gardens, where it makes its mark as a flowering tree of unusual beauty. It is well suited for the household bioplan for feeding and caring for a diversity of migratory species on the continent.

DESIGN

To both gardener and horticulturalist, *Asimina triloba,* the pawpaw, has many advantages. The tree will grow well in shade. There are very few deciduous, shade-loving small trees in North America that can fit well under a canopy and thrive to produce fruit, medicine, and pesticides. The fruit produces a rich, delicious pulp that can be used fresh in the kitchen or frozen for a later date.

The pawpaw grows well in a sunny, open location, where it assumes a pyramidal form. The lush dark green leaves have a prominent midrib, being up to 12 inches (30 cm) long and nearly 6 inches (15 cm) wide, tapering to fine tips. These droop, too, adding to the pyramid form. In the fall the leaves turn a bright yellow color.

The tree flowers on the previous year's wood, the flowers appearing in spring before the leaves. They are .5 inch (1.2 cm) wide, a beautiful, deep purple-black, and consist of two sets of three triangular petals opening around a central ovary. The flower has an understated beauty that is not too dissimilar to a heritage rose. The fruit clusters, when ripening, change from green to yellow to brown. Like the banana's, the outer skin of the fruit should not be eaten.

Asimina triloba can be underplanted with *Muscari commutatum,* a tiny, dark-flowered muscari, to repeat the tones of the pawpaw's flowers. This can be followed by the wonderful, black-flowering *Fritillaria persica* 'Adiyaman' to retain the beautiful black shades a little longer in the spring garden.

There are many interesting cultivars of *Asimina triloba.* The more superior are *A. t.* 'Middletown', *A. t.* 'Mitchell', *A. t.* 'NC-1', *A. t.* 'Overleese', *A. t.* 'PA-Golden', *A. t.* 'Sunflower', *A. t.* 'Taylor', *A. t.* 'Taytwo', *A. t.* 'Wells', *A. t.* 'Wilson'. Of these cultivars *A. t.* 'Overleese' has large fruit and *A. t.* 'Taylor' and *A. t.* 'Taytwo' hold the court for the finest flavor.

Asimina Species and Cultivars of Merit

SCIENTIFIC NAME	COMMON NAME	ZONES
Asimina triloba	Pawpaw	4–10
A. t. 'Carrigliath'	Carrigliath pawpaw	4–10
A. t. 'Middletown'	Middletown pawpaw	5–10
A. t. 'Mitchel'	Mitchel pawpaw	6–10
A. t. 'NC-1'	NC-1 pawpaw	6–10
A. t. 'Overleese'	Overleese pawpaw	5–10
A. t. 'PA-Golden'	Golden pawpaw	5–10
A. t. 'Taylor'	Taylor pawpaw	5–10
A. t. 'Taytwo'	Taytwo pawpaw	5–10
A. t. 'Wells'	Wells pawpaw	6–10
A. t. 'Wilson'	Wilson pawpaw	6–10

Oenothera brachycarpa, evening primrose. Hugging the Gulf Coast of North America, it is as heat loving as is *Asimina triloba*, pawpaw.

A mature birch and white pine wood. In time the pine will outlive the birch.

Betula

BIRCH

Betulaceae

Zones 2–9

THE GLOBAL GARDEN

The birch, together with the aspen and willow, started the present-day cycle of our modern forests. These trees were the first to move into the limelight after the last ice age of 12,000 B.C. To this day, birch grows in the Arctic in a dwarf or *nana* form, as it did at the time of the land bridge between North America and Siberia.

There were two sacred trees given to the aboriginal peoples of North America by the legendary Winabojo. These were the birch and the cedar. The Chippewa peoples in particular showed great ingenuity in their use of the birch. Great sheets of bark were gathered from living trees in June or early July. The bark was slipped off, collected, and stored. When the bark was needed for domestic items like cups, bowls, and kettles, it was heated and stretched into the form required. Birch was used to make cups, serving dishes, boxes, coffins, wigwams, *makuks,* cooking utensils, funnels and cones, meat bags, fans, torches, candles and tinder, dolls, and sleds. Birch made paper for writing and for art in a folded and stippled form known as transparencies. This was women's art only. This lost art is akin to quilt work of the white settlers of North America and similar to the old Celtic designs on woven lamb's wool quilts. It is also similar to the Japanese art form of origami. Birch was also used in complex medical practices that involved surgery and the setting of broken bones using look-alike limb braces strengthened with birch wood and stabilizers made from basswood twine.

Birch was fashioned into the famous birch bark canoes that echo in shape, form, and function the *curach,* or sea boats, of the seafaring, ancient Celts and present-day Aran islanders of western Ireland. Even moose calls made from birch rolls have recently been found in Mesolithic North American caves.

Historically we know that the birch tree became a universal tool for survival transcending continental boundaries. Its name was one of a handful of words, thirty-six in all, shared by the peoples who live in central Siberia and speak a language called Ket and the Athabaskan of the Na-Dene nation of Canada.

Betula, the birch, grows in the cooler, north temperate regions of zones 2–9. Birch is to be found in every northern country of the global garden. Because the birch is a short-lived tree, it has had time to hybridize and produce the species that are unique to their own particular turf. For that reason, there are approximately sixty species of birch worldwide.

Birch, be they the lady birch, *beith gheal,* of the Irish Killarney woods or the shining orange-brown *Betula albo-sinensis* var. *septentrionalis* of the Chinese forests, are all quite similar. They could be the fragrant medicinal trees, the cherry birch, *B. lenta,* of eastern America and its sister the Japanese cherry birch, *B. grossa* of mountainous areas. It could be the tall and proud yellow birch, *B. lutea,* around the Great Lakes, or the river birch, *B. nigra,* the only hot-blooded member, edging its way into Florida. Or it could be the last remnants of the five-thousand-year-old ancient birches of the Birch Province clinging to the tip of northern Scotland. These trees are all similar to the casual eye. The bark, the leaves, the growth pattern, the wintergreen fragrance in some cases, somehow bear witness to a common membership of the birch, or Betulaceae, family.

ORGANIC CARE

All birch can be grown from seed. The seed, or nut, is designed to travel great distances. It has two lateral wings and can float on the wind. The seed can even speed skate on ice. Occasionally this makes the collection of ripe seeds difficult.

Betula lenta, the cherry birch, a sweetly aromatic, native tree. Oil of wintergreen is the source of its definitive fragrance.

In the birch family both male and female flowers are produced on the same tree in the form of catkins. The male catkin is quite long and naked, the female catkin more squat and conelike in appearance. This cone has a tight, dense, compact feel when handled. This is the cone that is collected for seed. They are a greenish tan color when picked and can be dropped from the limb directly into a brown paper bag. They can be spread out to dry on newspapers in a warm area until they turn brown, at which time the seeds are mature inside the cone and a few will easily slip out. The cone can then be crushed over paper and the tiny dark seeds collected and stored for planting. The seeds can be either planted immediately or stored in brown paper in a refrigerator for later use. For long-term storage of up to two years, the seeds can be kept in an envelope at room temperature.

Birch seeds take around four weeks to sprout. The seed is unusual in the tree world because it needs light to help it germinate. As little soil as possible should be used to cover the seeds. The seeds can be placed ¼ inch (.6 cm) apart. They should be kept damp. When they germinate, the young cotyledons and embryonic leaves are helped to grow by a little shade. This can be supplied by one layer of Reemay placed over the seedbed or pots for a two-month period. It should then be removed and the seedling trees exposed to full light conditions.

The birch likes full sun and an acidic soil pH of 4.0–5.0. The soil should be rich and damp and be high in organic mater of low-nitrogen origin. For the birch this is, ideally, a leaf mold compost. These trees will not do well in shallow, alkaline soils. The birch can be easily transplanted as a potted or balled and burlapped tree up to 6 feet (2 m) in height. Many of the species birches are commonly available in nurseries, like the yellow birch, *Betula lutea* (syn. *B. alleghaniensis*); the paper birch, *B. papyrifera;* and the gray birch, *B. populifolia.* The European white birch, *B. pendula,* especially the *B. p.* 'Youngii', is increasingly used in North American gardens, as is the graceful, multicolored, Asiatic *B. ermanii,* the 'Grayswood Hill' birch.

The planting hole for birch should be large. As most garden soils are nearer to neutral, the soil mix going into the hole has to be acidified to match the needs of the birch to maintain its health. One gallon (3.8 l) of one- to two-year-old horse manure or leaf mold compost should be mixed with some forest soil and placed at the bottom of the hole. The latter helps the mycorrhizal fungi to become established around the young birch roots. To this is added two gallons (7.6 l) of peat moss and half a quart (.5 l) of dry woodash. These are mixed well with the soil in the hole. After the tree is planted, air pockets are stamped out by heel pressure, leaving a shallow area that will catch rain.

Global warming has a twofold effect on the health of the birch. The birch leaf miner, *Fenusa pusilla,* can cause greater casualties because earlier than normal broods will feast on succulent leaves in May. Antifeeding sprays like neem tree oil can help protect this foliage in a garden. The action of increased sulfur dioxide levels is also harmful to the stomata or breathing apparatus of the leaves. Positioning of the tree in the garden away from the route of automobile exhaust will maintain a healthier tree.

MEDICINE

The twigs, leaves, branches, and bark of the cherry birch, *B. lenta,* the Japanese cherry birch, *B. grossa,* the Virginia roundleaf birch, *B. uber,* and, to a lesser extent, the yellow birch, *B. lutea* (syn. *B. alleghaniensis*), carry wintergreen biochemicals important in their medical use. The white birch, *B. papyrifera,* and the gray birch, *B. populifolia,* are also occasionally used medically.

The following medicines are found in "fingerprint" amounts depending on the species of birch: betulin, betuloresinic acid, saponins, betuloside, gaultherin, methylsalicylate (an essential oil), and ascorbic acid (or vitamin C).

Biochemicals originally from the birch are to be found in almost every cupboard in every home in the Western world. One of these biochemicals is aspirin, or salicylic acid. Some fourteen million aspirin pills are consumed on a daily basis. Historically, the cherry birch, *B. lenta,* was steam-distilled to produce an oil called oil of wintergreen, which is high in salicylic acid occurring as the methylester. The famous Bayer company, a hundred years ago, introduced acetylsalicylic acid that still does a similar task today, as did the birch extract employed by North American aboriginal people for centuries. Specifically, it reduces fevers, stops pain, and acts as an anti-inflammatory agent. More recently, it is being used as a circulatory anticlotting biochemical that increases blood flow by reducing viscosity.

A tisane made from birch leaves was used in Russia, England, and the United States and by the aboriginal peoples of North America for the treatment of urinary tract infections. Fresh, mature leaves were gathered. A tisane was made from them using boiling well water. This tisane or tea was thought to have a gentle, mild, antiseptic action on the delicate tissues of the kidney and the urinary tract. It also functioned as a diuretic.

The aboriginal peoples of western North America used their only native birch, the water birch, *B. occidentalis,* as a common treatment for coughs, colds, flus, and other pulmonary ailments. A decoction was made of the mature bark. This was drunk, when needed, on a daily basis.

The Pillager aboriginal peoples collected the mature female cones of the swamp birch, *B. pumila.* These cones were dried and then were burned over hot coals to produce an aerosol incense that was used to relieve a catarrh of those suffering from pulmonary complaints. They also used a tisane made from the same cones as a postparturition tonic. This tea was also used for pain relief during difficult menses.

Twigs of *B. lenta,* the cherry birch, and *B. lutea,* the yellow birch, were used by the aboriginal peoples in the past and by the present inhabitants of the Ozarks in the United States. These twigs were used as chew sticks. The sticks were chewed and the softened ends were used much like a toothbrush, except that the medicine of the twig itself was extracted by the alkaline saliva to formulate a chemical block to the action of cariogenic bacteria of the teeth and mouth. Sometimes other chew twigs, especially of tropical origin, had a high fluorine content. This method of

Eupatorium purpureum, Joe-Pye weed, a native species found in birch woods. It was named for the medicine man Joe Pye, who used it to treat kidney stones.

oral hygiene has been practiced since the Babylonians in 2000 B.C.

The Mohawk aboriginal peoples used gray birch, *B. populifolia,* to treat bleeding piles. Seven strips of bark were taken from the east side of the tree. These were boiled in well water, cooled, and the decoction drunk as needed.

The birch produces a sugar called xylitol that appears to inhibit the growth of the bacteria, *Streptococcus pneumoniae,* responsible for tooth decay. This bacterium in its ordinary life hitches a ride on the water mucus of the mouth as it moves from throat to ears and back again via the eustachian tube. Here it begins the ear infections of childhood. Industry is now looking at a chewing gum with xylitol as a sweetener. This birch sugar will have a twofold function of clearing ear infections and maintaining healthy teeth, particularly in the older mouth.

Many medicated soaps are made from oil of wintergreen. This soap can be tolerated by people suffering from sensitive skin and various other skin conditions.

Medicated hard candies called wintergreen candies were used by the early pioneers as a means of preventive medication for colds, flus, and other bronchial ailments contracted during the long, cold winter months.

ECOFUNCTION

Both large and small, the birch is a tree of forest regeneration. It can be considered to be the nursery ground for the long-lived trees in the wildwood. This is because the birch has a remarkable ability to seed into either burned or cut-over areas. The seeds surface, sprout, and live just long enough, 100 to 150 years, for the next generation of succeeding maples, beech, basswood, eastern hemlock, balsam fir, eastern white pine, oaks, and both the white and red spruces. The oaks, pines, and maples, especially, receive enough shade to get established. This is a very important function in forest regeneration.

Many species of birch, such as the water birch, *B. occidentalis;* the river birch, *B. nigra;* the cherry birch, *B. lenta;* and the Virginia roundleaf birch, *B. uber;* track streams and flowing water. These species are all important for bank stabilization of lakes and rivers. These, too, are the species that produce wintergreen biochemicals. It is possible that these biochemicals have a protective function for fish populations and spawning beds.

In particular, the cherry birch, *B. lenta,* is considered to be a deer medicine by the Seneca peoples. This is because the plant sustains health in the deer herd and seems to be attractive to them as winter browse. The buds of all species of birch are eaten by overwintering large birds such as partridge, grouse, and turkeys. Rabbits and hares also eat the twigs in winter, as do mountain sheep and goats whenever possible.

The birch, because of its ubiquitous nature, is an important, global host tree for butterfly populations. The chemical protective load in the leaves confers an avian protection on these butterflies. The tiger swallowtail, *Pterourus glaucus;* the green comma, *Polygonia faunus;* the Crompton tortoiseshell, *Nymphalis van-album;* the white admiral, *Basilarchia arthemis;* the honeydew-drinking butterfly and the dreamy dusky wing, *Erynnis icelus;* will use the leaves of all the North American birch species.

Many of the birch species have domatal hairs on the undersides of their leaves that are repositories of small, predaceous insects waiting to hear their terpene call to aid in some attack nearby. It seems that many plant species being eaten by caterpillars will go through a certain sequence of action. The leaf will taste and be aware of the chemistry of the saliva of the dining insect. If necessary, the leaf will then secrete a terpene, which is highly volatile like a perfume, into the air. The predaceous insects who are waiting in the domatal wings will smell the delicious aroma and jump into the fray. The diner becomes the dinner. The bottom line is survival, and the tree usually has the upper hand. The cycle has to remain that way for a forest ecosystem to survive.

Many species of birch are directly associated with certain mushrooms: the gray birch, *B. populifolia,* with *Boletus scaber;*

the white birch, *B. papyrifera,* with *Entoloma jubatum;* the yellow birch, *B. lutea,* with *Pholiota luteofolia.* The mycelium or asexual growing apparatus of these mushrooms forms the arbuscular, mycorrhizal growth associated with birch roots. Failure of this symbiosis spells death either for the tree or the fungus. These associations are not understood, rarely studied, and mostly ignored.

BIOPLAN

The cherry birch, *B. lenta,* is native to Canada. This has been proven by paleoecologists, who can determine from the natural architecture of root growth whether a tree has been hand planted or if it is, in fact, native. There are few examples of this tree left. These are to be found growing on the shores of Lake Ontario near Port Dalhousie. These few northern specimens will, of all of the birches, have the highest medicinal action. These species should be protected and propagated as part of Canada's living heritage.

A second birch is listed as officially endangered, the Virginia roundleaf birch, *B. uber.* This species is found growing in Smyth county of southwest Virginia. This species, too, has significant medicines that will be lost if this birch is not propagated.

Both *B. uber,* the Virginia roundleaf birch, and *B. lenta,* the cherry birch, could be the source of a cottage industry. Both these trees produce an oil that is strongly fungicidal. If this oil is used to treat leather, it prevents the growth of mildew and other leather-eating fungi. It could be used in the conservation of leather-bound books for libraries and for the antique book trade. Birch oil conditioner could also be of use in the equine world for the long-term storage of saddles, harness, and bridles in tack rooms where they are prone to mildews.

All of the birches, especially the yellow birch, *B. lutea,* can be tapped in the spring as a source of birch sap. This sap can be reduced to make birch syrup, which is not unlike maple syrup but is a health food, being high in xylitol. The sap can also be

Gaultheria procumbens, wintergreen, the ancient and original source of wintergreen as an aboriginal medicine for the treatment of colds. It is a native companion to the birch.

fermented into a health beer, a delicious wine, or a northern vinegar. The inner bark can be collected and dried into a flour that, when added to other grain flours, makes a high-protein bread that is naturally sweet.

Birch sap had been used in the past in cosmetic manufacture because its astringent action is gentle on the skin. Of the many species of North American birches, both *B. uber,* the Virginia roundleaf birch, and *B. lenta,* the cherry birch, would make superior cosmetics in addition to medical soaps, shampoos, hair rinses, and balm for chapped lips.

The yellow birch, *B. lutea,* is a native of the north temperate forest. Over the past 50 years it has become scarce. In fact, there are few yellow birch to be found over 100 years old. Where they are growing, they do not appear to be regenerating. Within a known span of 50 years of growth, no new seedlings have been recorded by the author in a local forest. For this reason alone, the yellow birch should be added to reforestation projects and especially planted in damp, shaded areas where these trees can have cool roots to withstand the summer temperature fluctuations of the future. The yellow birch is particularly excellent as a species for reconstruction of riparian areas.

All parts of all of the *Betula* species can be reduced to charcoal and a tar product that can be used for caulking boats.

Betula lenta, the cherry birch of Canadian origin; *B. grossa,* the Japanese cherry birch; *B. uber,* the Virginia roundleaf birch, *B. lutea,* the yellow birch; and *B. utilis* var. *occidentalis,* the Himalayan birch; can be bioplanned into a hospital garden. The sitting area should be exposed due south in the warmest position. A mixture of these birch can be used to the north side of the seating area. Leaves, twigs, and cones should be collected and maintained as a mulch in this small garden. The aerosols released from this grouping will benefit patients recovering from prostate cancers, kidney dialysis, and kidney transplants. It will benefit those who need any repair of the urogenital areas.

DESIGN

The paper birch, *B. papyrifera,* has long been popular as a specimen tree in Canadian gardens. The European white birch, *B. pendula,* with its graceful, weeping tiers of branches, is slowly creeping into city gardens as *B. p.* 'Youngii'. This delightful tree is tougher and hardier than was previously thought. There is also another cultivar, *B. p.* 'Purpurea', which is a purple-leafed weeping birch. The European white birch and its cultivars make splendid trees for smaller gardens. This tree puts a photo finish on any small garden, giving it form for all seasons. All this tree needs is a massing of snowdrops, *Leucojum* species, at its feet. This can be followed later on in June by an irregularly shaped bed of one of North America's best native bulbs, the camas, *Camassia leichtlinii* 'Alba', whose tall white spires accentuate the white bark of the weeping birch. Both bulb and tree enjoy damp feet. The bulbs disappear quickly into the lawn.

No serious herb garden or medicine border would be complete without the cherry birch, *B. lenta.* A design to fortify this tree would be the use of birch maypoles within the border or as part of the herb garden, festooned with colorful ribbons, as a counterbalance to the birch. These trees in the Western world have long been connected with magic and pre-Christian fertility rites in the Northern Hemisphere. Their positioning in a border is ethereal and gives the custom of the birch its correct salutation.

Betula Species and Cultivars of Merit

SCIENTIFIC NAME	COMMON NAME	ZONES
Betula albo-sinensis var. *septentrionalis*	Chinese birch	4–8
B. ermanii	Grayswood birch	4–9
B. grossa	Japanese sweet birch	4–8
B. lenta	Cherry birch	4–8
B. lutea (syn. *B. alleghaniensis*)	Yellow birch	3–8
B. nigra	River birch	4–9
B. papyrifera	Paper birch	2–8
B. pendula	European birch	4–9
B. p. 'Purpurea'	Purple European birch	4–9
B. p. 'Youngii'	Young's European birch	4–9
B. populifolia	Gray birch	3–8
B. pumila	Swamp birch	2–8
B. uber	Virginia roundleaf birch	4–8
B. utilis	Himalayan birch	3–8

Camassia leichtlinii 'Alba', the white camas, one of North America's best native bulbs. It accentuates the white bark of the birch.

The exfoliation of huge scales of bark of *Carya ovata*, the shagbark hickory, forming a river into the sky

Carya

HICKORY

Juglandaceae

Zones 3–10

THE GLOBAL GARDEN

Few who have tasted the delights of butter pecan ice cream look to the hickory tree with gratitude. The pecan is one of the hickories. The hickory is an American child, blind to boundaries, living on the face of a continent in global solitude. Despite the millennia of growth and sustainable care given to these trees by the native aboriginal peoples, there is scant oral history remaining to tell us of the hickory's unique heritage.

The word *hickory* is of aboriginal origin, *pohickery* or *pawhiccori,* dropped to hickory. Even *pecan* is derived from the Algonquin *pakan* or *pagann.* In any case, these were the botanical terms the first settlers had heard to identify the enormous nut-bearing trees they saw. These trees were entirely new to them. The few remaining suckering boles indicate that the hickories had circumferences of 18–20 feet, or 5–6 meters, even in poor soils!

When these settlers arrived, had they paid less attention to cloutie dumplings and more attention to the culture of the hickory, North American agriculture would have a different bent today. Hickory smoke would be to the Canadians as the grape is to the French.

The saw is as silent as the sword, for relentless cutting of the estimated one billion board feet of standing *Carya cordiformis,* bitternut hickory, to the annual shearing of 300 million board feet of *C. laciniosa,* the kingnut hickory, has diminished a sustainable agro-society to what it is now. The few forest refugees we have left are just a fingerprint of what was there before. These serve only to pinpoint an ancient habitat. They also are a further testimonial to the horrors of the British penal laws that shaped the settlers' minds because fire and famine was all they had known and would know if the virgin forests were not felled. It was the law of possession of their adoptive land. They had to clear their lands for the right to own them. Their country had changed, but the masters remained the same.

The hickory is a strange breed of tree, shifting its sights with heat and with cold. It grew in much of Europe before the last ice age. The encroaching ice extinguished an entire race of trees. Although hickory will grow in England, it will only occasionally produce a crop of nuts in an exceptionally long hot summer. In Brazil, the hickory appears to explode with growth and enormous crops of nuts.

Species of hickories spill out into eastern North America. The bitternut hickory, *C. cordiformis,* is found growing on the poorer, higher land. The pignuts, *C. glabra,* will take the better soil of mountainous areas. The shagbarks and shellbarks, *C. ovata* and *C. laciniosa,* both grab the deep loams of fields. The mighty pecan, *C. illinoensis,* veers for the top of the heap in soils. They swank out in the deepest, richest, alluvial, river bottom soils, stretching their 170 feet (52 m) in height over the finest of land from Alabama to Missouri and, more recently to upstate New York and Ontario. There is also an aquatic member of the team, the water hickory, *C. aquatica.* There are some dry-landers, the sand and black hickories, *C. pallida* and *C. texana.* Related to these desert camels is the one Mexican species, *C. palmeri.* After the continental divide Southeast Asia kept two hickories, *C. tonkinensis* and *C. poilanei.* And lurking in the Chinese forests is the tooth-smashing 'Iron' nut, *C. sinensis.* But very little is known about this potential dodo of the hickory world.

Probably because they are wind pollinated, hickories have flexible genetics. Over time the hickory seems to have made a

Mature, edible nuts of *Carya ovata,* shagbark hickory, waiting to be picked by man or beast

tetraploid of itself, primarily trying to hang on tooth and nail to the tougher regions of the continent. The chromosome number of 32 doubled to 64. This genetic blessing has given rise to a great number of natural hybrids. These hybrids will be expressed in the nut through size, shape, meat, ability to crack, ability to slip easily from the shell, and in timber characteristics of the tree.

The hickory produces the most delectable nuts in the world. The fresh meats have a crunchy, creamy flavor that seems to change to butter when they are cooked. In all of the nuts there is very little bitter aftertaste, just a lingering of sweetness and a pecanlike, nutty flavor on the tongue. The taste of each crop of nuts is unique to its species and somewhat to the season's growing conditions. Topping the list for quality are the pecans, the shagbarks, and the kingnuts.

ORGANIC CARE

Every single hickory species can be grown from its nut. The nuts are collected early in the fall as soon as they begin to mature. On the tree the nut is seen to be green. This is the outer, soft husk that surrounds the hard nut at the center. For all of the hickory species, the husk opens into quadrants starting from the tip. All of the hickory trees produce warning signals for nut collection. Immediately after the so-called floaters fall, the tree is open for picking. For some reason, the hickories produce an early flow of a hormone called abscissic acid to abort the insect or fungally injured nuts first before the good nuts are ready to drop. The chemistry of this alerts the squirrel army who make a living out of hickory nuts. It should also be a clue to the intrepid human picker that the chase is on.

Hickories exhibit some embryo dormancy in the nuts. What this means is that the nuts are not yet fully mature to germinate. The dormancy changes on the face of the continent. In growing zones 7–10, it is minimal. It can be up to 30 days. For the colder growing zones, 3–6, this dormancy is longer and can be from 60 days up to 150 days, the colder the zone the longer the time. Some hickories at the edge of their natural habitat have a very long internal clock for dormancy. This can be up to two years. This is seen in the kingnuts, *Carya laciniosa,* in particular. Dormant embryos are changed into growing embryos by a waiting-game process called stratification. For stratification, the nuts should be dehusked and stored at 41°F (5°C) either in a refrigerator or a cold storage room with good humidity. This can be done in the home by placing the nuts in a clean glass jar and putting them in the back of the refrigerator for the given period. The jar should be dated.

In the spring, after the stratification time is over, the nuts can be placed outside. The nuts are planted on their bellies, 2 inches (5 cm) deep and 6 inches (15 cm) apart. They should be transplanted to their final site the first spring following germination as one-year-old seedlings.

The nuts may also be potted up indoors in a fifty-fifty sand and soil mixture. This is kept damp but not wet. The pots are then subjected to 86–90°F (30–32°C) daytime temperatures, dropping to nighttime temperatures of 68–70°F (20–22°C). These pots will benefit from a light layer of wood chip mulch if it is available. Once germinated, the young seedlings can be carefully planted outside.

Increased solar radiation in zones 7–10, and cold variations in zones 3–4, will affect the growth potential of the young hickory seedling. This can be bypassed by the use of Reemay. The seedbed or seedlings can be draped with one layer of this cloth. In zones 7–10, it will reduce the damaging effects of a high solar exposure in early spring. In the colder zones it will give 9°F (5°C) of frost protection to the young seedlings.

Once germinated, the young hickory seedling grows rapidly. The seedling has its own sun screen. This is seen in the reddish color of the elongating tree. This hue changes when the growth phase is complete and when the first apical meristem is formed and matures. Young hickories are very tough and enduring. In the colder zones 3–6, hickories break dormancy late, usually toward the end of June.

Hickories should be transplanted in their first or second year of growth. The stem might be only 6 inches (15 cm), but the roots, even at that young stage, are considerable. The soil should be made as rich as possible, with the addition of aged bonemeal or soup bones at the base of the hole as a slow-release phosphate fertilizer. Liberal amounts of aged horse manure or compost should also be added and mixed well with the soil in the hole. A source of calcium is necessary for all nut-bearing trees. This can be added in the form of finely ground dolomitic limestone to the planting hole. Limestone and woodash can also be applied, as the years go by, around the tree to the drip line. This provides a slow release of calcium and helps combat the acidification effects of atmospheric pollution. It can be applied once in every five years. The soil pH for hickories should, however, remain a little on the acidic side, 6–6.5. But they will also grow very well on alkaline soils. The shagbark, *C. ovata,* and the bitternut, *C. cordiformis,* thrive with an underburden of limestone.

Mature hickory trees should be pruned in the fall when the tree is fully dormant. This will vary with the growing zone. In the colder zones, the months of November and December are safe. In years of lunar perigee, fall pruning, for all zones, should be kept to an absolute minimum. Hickories should always be pruned aseptically. The cut should be made and the area swabbed with either alcohol or a 2 percent household chlorine bleach solution, 1 ounce bleach to one quart water (20ml/l). The cutting tools should also be sterilized in between cuts. A sterile paper swab can be left on the tree. This seems to promote very rapid callus healing the following spring.

Hickories do not transplant readily. This is the reason they are so rarely available in nurseries or forest stations. In any case, the younger the tree, the better the take. So those with a little gardening spirit in their souls will do very well with this tree species.

Every hickory tree is self-fertile. What this means is that every tree has the capacity to bear a crop of nuts. Both male and female flowers occur on the same tree. The male pollen is wind-distributed and has an excellent chance of fertilizing the female flower. But the life of the hickory is not an easy path. These giants of the forest also have internal chronometers with the same circadian chemicals found in humans. So, depending on an internal decision made by the tree, pollen is sometimes liberated a little after the female flowers peak. In some years this means the nut crop will be poor and in other years it will be heavy. This phenomenon of peaks and troughs of productivity is a complex one and is deeply tied into the tree's philosophy for survival. It is at this point where cross-pollination from other related or genetic hybrids with chromosomal multiples of two come in handy to increase a nut crop in any given year in a hickory plantation or producing nut grove.

From time to time there are a number of disfiguring cankers and leaf spot diseases that attack the hickory population. These are mostly fungi. Humidity in summer also brings its share of powdery mildew active on the leaf surfaces. These can be kept at bay with the following practices. As the ground warms up in the spring, dry woodash can be dusted to coat the entire trunk up to a height of 3 feet (.9 m). Particular attention should be given to the collar where the trunk meets the soil. The pH of the woodash application is such that the fungal spores are killed before they

The unmistakable bud of a shagbark hickory, *Carya ovata*

can infect the tree. In addition, the ash puts a chemical block to discourage ants who may want to farm aphids higher up the tree.

For the home gardener, farmer, or those interested in the commercial aspects of beginning a hickory plantation there are a number and combination of exciting choices available. Popular pecan cultivars include *Carya illinoensis* 'Earlibest', *C. i.* 'Posey', *C. i.* 'Starking'. Its various frost-hardy hybrids with the shagbark hickory are known as hicans. The finer nuts are the Burton nut, *C. ovata* × *C. illinoensis* 'Burton' and the Des Moines nut, *C. ovata* × *C. illinoensis* 'Des Moines'. Both the shagbark hickory, *C. ovata,* and the kingnut hickory, *C. laciniosa,* have superior nut cultivars. The *C. o.* 'Neilson', *C. o.* 'Yoder #1', *C. o.* 'Walters', *C. o.* 'Weschcke', and *C. o.* 'CES-26' bear fine nuts, and *C. o.* 'Carrigliath' fine-tunes the shagbark into bearing nuts in the colder zones of 3–4. The larger kingnuts, *C. l.* 'Henry', *C. l.* 'CES-24', *C. l.* 'Fayette', 'CES-1' and *C. l.* 'Hoffeditz' need a higher potassium-phosphate ratio in the soil to give the trees a little extra cellular frost protection. The sweet nut forms of *C. cordiformis,* otherwise known as the bitternut hickory, have growing zones of 3–10. These small nuts are delicious and make a high-grade oil for cooking, eating, or lighting. There are many natural cultivars of these bitter nuts growing in the wild yet to be discovered and propagated. For cross-pollination *C. tomentosa* and *C. glabra* 'Brackett', the mockernut and the pignut hickories, add compatible pollen into a plantation of nuts. This can cause wild crosses also for experimentation.

At the moment the hickory seems a poor candidate for cross-grafting within its own family. Over time, graft incompatibility arises. This and all aspects of nut culture are being explored and researched by a very worthy and noble organization, the Northern Nut Growers Association.

MEDICINE

The health of the hickory is found in the nut and the medicine in the tree. The nuts are high in fatty acids. These are called oleic, linoleic, and linolenic acids. They are essential fats, being absolutely necessary for brain development and healthy organ functioning in man. A lack of them gives rise to a range of diseases, some of them life threatening, while a reduction of them delays repair in the body.

To the aboriginal peoples of North America, the hickory family was an important one. These trees were tended by biannual flash firing, a horticultural technique to increase the nut crop.

The shagbark, *C. ovata,* and the kingnut, *C. laciniosa,* were used interchangeably. They also actively searched out the sweeter forms of the bitternut, *C. cordiformis,* and pignut, *C. glabra,* on a continuing basis.

The Mohawk peoples used the bark from either the shagbark, *C. ovata,* or the kingnut, *C. laciniosa,* as a treatment for intestinal worms. Hickory bark together with balsam poplar, *Populus balsamifera,* and butternut, *Juglans cinerea,* barks were taken from small branches of a minimum of 3 inches (8 cm) in diameter. The inside bark was used. This bark is the white xylem-wood area. A single handful of poplar and hickory was scraped off. This was mixed with a double handful of bitternut bark. The combined barks were put into 4 pints (1.9 l) of well water. This was boiled down to 1 pint (.5 l). An equal volume of sugar was added to make a syrup. The syrup was taken twice a day before meals in the morning and in the evening. Following a bowel movement it was repeated after an interval of four days until the system was worm free.

The Cayuga used either the shagbark, *C. ovata,* or the kingnut, *C. laciniosa,* as a remedy for arthritis. Two bark bundles or handfuls were taken in early spring from the east side of the tree. Both bundles were kept separated and boiled in well water. From one sample, the decoction was used as an external poultice directly on the limb. The other was drunk as an internal medicine. Both internal and external medicines were used together to relieve the symptoms of arthritis.

An oil was extracted from the bitternut hickory, *C. cordiformis.* This oil was very useful to the aboriginal peoples in the treatment of rheumatism. This cold-pressed oil probably has quercetin and quercitrin, the rhamnoside glycoside. These compounds have a direct action on circulation with entry through the skin. The effect is local on the capillaries and thus helps ease the pain of rheumatism. It is especially useful when the oil is applied externally as a rub. This type of practical medicine is very old.

A health food not unlike tofu was made from the nuts of the shagbark hickory, *C. ovata,* and the kingnut hickory, *C. laciniosa.* A nut cream was extracted using hot water separation. The nut cream was collected and concentrated. It had excellent keeping properties when kept cool. This high-protein, high-fat food was stored and used as an enricher in all manner of additions to the aboriginal cuisine. It was said to be particularly good with duck or wild yams, *Dioscorea villosa.* It was added to soups and stews and used with other flavorings. A nut cream was also made only from shagbark nuts, *C. ovata.* This cream was allowed to ferment. The fermented product produced a mildly alcoholic winter beverage called *pawcohickora.*

The green nuts of the shagbark, *C. ovata,* and the kingnut, *C. laciniosa,* were collected when green and just ripe. The dehusked nuts were crushed to expose the nut meats. They were then thrown into water that was at a rolling boil. The cream rose to the top. This was skimmed off with a ladle. The cream was collected and further concentrated down by heat. This product was then kept sterile. The nut cream kept exceedingly well.

The sweet nut meats of the hickories were also air dried. These were ground into a nut flour. This high-protein flour was added to other flours to make a high-energy bread. The hickory nuts and other nut crops were stored in special tripod buildings that allowed air movement through the nuts. This process of storage made the nut meat separate easily from the surrounding shell.

A cold-pressed oil was also extracted from the bitternut hickory, *C. cordiformis,* that was similar to virgin olive oil. This was used for medicine, lighting, and cooking. The early white settlers "borrowed" this oil for their lamps.

Chewing gum has been a North American aboriginal custom for millennia. This is a tooth-cleansing process helped by the oil

Carya ovata, the shagbark hickory. It has a stunning, shell pink color in the spring when the growing tips decide to elongate.

Carya cordiformis, bitternut hickory, which shows a fingerprint of the past. The huge 18' (5.5 m) bole of the virgin tree forms a clone of trunks from the original bole's roots. Many of these remain in North American forests.

in the bark of hickories. The fresh bark of many species of hickory was used as a chewing gum.

ECOFUNCTION

Hickory wood is extremely dense and heavy. It is also very shatterproof. This is the reason why it is used in high-quality, trail-breaking cross-country skis and ax handles. The extreme density of the wood gives it very special qualities of durability. It also means that all the hickories, *Carya* species, need huge amounts of carbon dioxide to grow because carbon dioxide is the building block material that makes up collenchyma and schlerenchyma tissues. These fibers are packed tightly together to give the wood its density. They are like carbon pumps packing carbon into every chemical nook and cranny of the tree, so much so that it is difficult to read the annular rings of growth from spring to fall.

As the *Carya* species grow, this carbon bank doesn't end there. All the hickories have underground associations with culturally advanced fungi. These fungi are like underground cities with factories that run on carbon from carbon dioxide. The trees are the carbon trading partners of these invisible fungi in a commerce that begins with atmospheric carbon dioxide. Massive, wholesale logging of these trees in their north temperate deciduous forests considerably adds to global warming. The trees act as a carbon sink. Break the sink and you flood the atmosphere. The cutting of global forests smashes the sink, the plumbing, and the drains. It is thought that the cutting of the global forests is responsible for 25 to 30 percent of global warming. These figures might be higher because the carbon traffic from the trees into the living soil is not, as yet, understood by scientists.

The planting of hickories, with *Carya tomentosa*, the mockernut, topping the list, followed by all of the other hickory species will help reverse global warming. This is the ecofunction for these trees in the twenty-first century, the reduction of atmospheric carbon dioxide. There is nothing more important than this, the return of the forests to human life on this planet, because forests clean the air of carbon dioxide and pump out oxygen for all living systems. Kill the forests and aerobic diversity goes, including mankind.

In the forest, the hickory is a feeding tree. They feed squirrels and the larger mammals. They feed songbirds and particularly the larger birds like grouse and wild turkeys, and also butterflies and insect populations. Historically the mast of the forest has been used to feed pigs.

Trees that feed squirrels help to expand a forest, for squirrels are nature's foresters. They have a nasal ability of selection; the diseased nuts are discarded and the good nuts planted. An increase of squirrels brings an increase of other wildlife, even birds.

All of the hickories are host plants for the banded hairstreak, *Satyrium calanus*. These butterflies engage in territorial tussles in which a dozen or so will chase one another around before alighting on a milkweed to drink. The hickory hairstreak, *S. caryaevorus,* also uses the hickory species as a host. These can sometimes be seen in clouds of the more common banded hairstreak butterflies.

The roots of the hickories are unusual. The mycorrhizal fungi that associate with them form a feeding sheath around the root itself. This sheath is composed of a complex network of mycelia, which are elongated, working strands of the fungi. The sheath over the apical tips forms a caplike structure called a hartig net. But what is so interesting is that the fungi that form this feeding hartig net are from the higher-order fungi. There are about 5,000 or more fungal-mycorrhizal combinations possible for each hickory species. These fungi are the trade workers of a living soil. They weave their lives invisibly underground only to pop up a mushroom body as a sexual expression of change for reproduction. This can happen once a year or once in a hundred years when a genetic marriage is needed for reasons that are not understood. The daily routine reproduction of these fungi is asexual or self-cloning. Scientists do not fully understand these phenomena.

BIOPLAN

North American hickories are species for forest plantations and countryside projects. They are not good trees for urban forests, homes, schools, and other institutional building landscape design. The reason for this are allergies. Hickories produce enormous quantities of wind-borne pollen in the spring. This pollen itself has a unique architectural form that can increase allergic reactions. On a global scale allergies are on the rise due to the particulate matter in the air from pollution.

The hickory forests should be replanted to combat global warming. This can be done in both Canada and the United States as part of the global effort to buffer the creep of surface temperatures.

Hickories can be bioplanned into the farming community as a cash crop for the fresh nut and salted nut market. The nuts can also be a source of cold-pressed nut oil, or nut cream and nut tofu. A fermented hickory drink and nut liqueur can also be produced from hickory nuts. These trees can be planted as part of an existing hedgerow or can be planted as a hickory savannah covering up to 20 percent of a field surface. Both of these bioplans take the form of two-tier agriculture where the nut crop of the trees is harvested as a cash crop in addition to the normal agricultural cash flow. The hickories, in turn, protect the farm animals from cancerous melanomas caused by the increases of ultraviolet radiation in the atmosphere.

Hickory plantations can also supply the exotic lumber market. The sawdust from this market can supply a source of hot or cold hickory smoke, the best being *C. tomentosa,* the mockernut, for a cottage industry of smoking hams, fish, and cheeses. All of these products could have a place in the local specialty markets that, in turn, help the tourist industry.

But the hickories of North America could even have a brighter future based on a new mushroom industry. This localized industry is presently in the Mediterranean area and is being slowed down by global warming there. The root tips of *Castanea* species or the sweet chestnut have hartig nets. In Spain these species are

The delightful, large-flowered *Uvularia grandiflora,* bellwort, a spring companion to all the *Carya* species

Tiny bristle ferns, *Trichomanes boschianum.* They are being lost by logging worldwide. They are associated with the high canopy provided by ancient *Carya* species. They are also found in midaltitude cloud forests and in one remote area of Killarney, County Kerry, Ireland.

under the microscope to see if they could produce crops of gourmet mushrooms like truffles. If this research succeeds, then the hickory species are next in line. The global demand for these gourmet mushrooms is not being met by present wild species markets. It is a very lucrative agricultural industry.

It is interesting to note that at present there is not a single standing collection of all of the hickory species and their cultivars in any public arboretum in Canada or in the world.

DESIGN

The most beloved of the hickories in the north are the shagbark hickory, *C. ovata,* and the kingnut or shellbark hickory, *C. laciniosa.* In the south are the majestic pecans, *C. illinoensis.*

At around fifteen years of age, right at the point of fruiting, the bark of the *C. ovata* begins to curl outwards and then peel. This is called exfoliation. The bark peels into long shaggy plates that are mostly free at both ends, especially at maturity, giving the tree a delightful shaggy dog appearance that is most unusual in the garden or forest. The trunk form is unforgettable.

The spring growth of both shagbark and shellbark is a shell pink color that is almost translucent. These rich tones are repeated in the male flower clusters donning the entire tree with a beautiful fragile appearance. In the garden the pink tone can be carried into the daffodil world using the larger pink daffodils such as *Narcissus* 'Louise de Coligny', *N.* 'Mrs. R. O. Backhouse' and the particularly beautiful *N.* 'Articole', which flowers a little later than the other two. These daffodils can be mass planted around the hickories to give a most pleasing effect in a large estate, a farm, or a country garden.

All of the hickories produce a light, moving dappled shade. This is because the trunk of the tree is long and the large leaves are pinnate. During the summer these leaves are a soft green that turns to a golden blend in the fall. The shellbark hickory, *C. laciniosa,* in addition, has butter-colored apical tips and the bitternut, *C. cordiformis,* has bright lemon tips.

The hickories make an excellent addition to a nuttery. Each species of the hickory family will produce its own crop of nuts. All of the species make excellent candidates for nut walks or nut allées.

Carya Species and Cultivars of Merit

SCIENTIFIC NAME	COMMON NAME	ZONES
Carya cordiformis	Bitternut hickory	4–9
C. glabra 'Brackett'	Brackett pignut hickory	4–10
C. illinoensis	Pecan	6–10
C. i. 'Early Best'	Early Best pecan	6–10
C. i. 'Posey'	Posey pecan	6–10
C. i. 'Starking'	Starking pecan	6–10
C. laciniosa 'Henry'	Henry kingnut	4–10
C. l. 'CES-1'	CES-1 kingnut	4–10
C. l. 'CES-24'	CES-24 kingnut	4–10
C. l. 'Fayette'	Fayette kingnut	4–10
C. l. 'Hoffeditz'	Hoffeditz kingnut	4–10
C. ovata 'Carrigliath'	Carrigliath shagbark	3–10
C. ovata × *C. illinoensis* 'Burton'	Burton hican	4–10
C. ovata × *C. illinoensis* 'Des Moines'	Des Moines hican	4–10
C. o. 'CES-26'	CES-26 shagbark	4–10
C. o. 'Neilson'	Neilson shagbark	4–10
C. o. 'Yoder #1'	Yoder #1 shagbark	4–10
C. o. 'Walters'	Walters shagbark	4–10
C. o. 'Weschcke'	Weschcke shagbark	4–10
C. o. var. *fraxinifolia*	Kingston shagbark	4–10
C. tomentosa	Mockernut hickory	4–9

The result of careless cutting, which caused the massive pilei of the *Steccherinum septentrionale* fungus to fruit. The fruit has the strange odor of ham as it dries down into the winter. It will hollow out the tree in time.

Airy billows of flowers, which occur in loose fragrant clusters. This is the hallmark of the *Catalpa.*

Catalpa

CATALPA, INDIAN BEAN TREE

Bignoniaceae Zones 3–9

THE GLOBAL GARDEN

Catalpa speciosa is the most spectacular summer-flowering tree of the north temperate regions of the world. The first half of July is greeted with airy billows of flowers in loose, fragrant clusters. Each individual flower could be said to be like a foxglove's, but has the ridging and form of a showy tropical orchid. The chaste whiteness of the petals is splashed with mauve to purple and gleams with raised golden bars. As the flower reaches fertilization, a carillon of white bells carpets the ground under the tree.

C. speciosa is just one of a small genus of 11 trees of various sizes, the principal two being found in North America. The remaining evergreen members lurk in the West Indies and eastern Asia. The most dominant tree with the largest flowers and fruits is *Catalpa speciosa*. There is the smaller southern cousin, *C. bignonioides*. The Chinese relative is *C. ovata*, not quite as beautiful as the *C. speciosa*, mainly because there are less flowers, but it is, nonetheless, useful in European gardens. *C. ovata* blooms in July and sometimes in August depending on the latitude. There is also a smaller second cousin, *C. fargesii*, which is more petite in every way. Not to be beaten for its perfect umbrella form is the top-grafted dwarf or *nana* sport of *C. bignonioides*. When this is grafted on to another catalpa species, it gives the delightful, standard, green mushroom form found so often in an Irish garden.

Catalpa speciosa in late June. The giant tropical leaves can be used as a soft bandage.

It is thought, by some, that native North American wildwood catalpas originated around the juncture of the Ohio and Mississippi Rivers. The knowledge of the exact habitat has been lost. This tropical tree has extended its range northward over the last century or so. It has become a viable, self-sustaining species in the wild with frost-resistant characteristics and the ability to self-seed in many areas of southern Canada.

The North American aboriginal peoples named the species, calling it *Kuthlapa*, meaning "head with wings." There is also a curious naming from the ancient Gaelic *Ceann leaba*, which means head rising out of bed. Both of these names are apt for the catalpa, showing an understanding of the tree's botanical origin from seed. As the seedling arises out of the newly planted seed, it appears to be a sleepy mop-head arising by the shoulders from the mattress of the ground. This head appears at the dead center of the seed, and the proportions are definitely human.

The Cree, the Cherokee, and the other First Nations were aware of the medicines of the catalpa. The present-day medicine women maintain that this species was lost by indiscriminate cutting and onsite burning of a wood that was soft, light and not very strong, which was the opposite of what was needed by the

Internal flower color guides insects for feeding in all the *Catalpa* species.

pioneer community. By contrast the aboriginal peoples had historically taken the long view of nature and were always thankful for the abundance of diversity surrounding them.

All of the catalpas are members of the trumpet creeper, or Bignoniaceae, family. In the fall each of the 11 species produce long beans that hang from the trees in vertical cascades. The long beans look like drying cigars. This has given rise to the catalpa's common name. Both North American catalpas, *C. speciosa* and *C. bignonioides,* are called Indian bean trees. The *C. speciosa* is also called the northern catalpa, the hardy catalpa, catawba, western catalpa, western catalpa tree, and the orchid tree.

ORGANIC CARE

Catalpa can be grown all over North America in zones 3–9 as an ornamental tree. The author has, over the past 25 years, bred a very frost-hardy cultivar, *C. speciosa* 'Carrigliath', in open field trials. This cultivar extends the growing range of catalpa into zone 3. This cultivar has also been selected to have additional xerophytic characteristics that makes the tree able to withstand increased cold, drought, and ice.

North America, with its very high solar conditions, is ideally suited to the growth of this splendid tree. Even though catalpas grow in European gardens, nothing prepares the gardener for the sight of *Catalpa speciosa* in full bloom in a Canadian garden. In full sun, when used as an open grown specimen, the catalpa reaches its finest flowering form. It will bloom in some shade, where it is inclined to have a more vertical habit.

Catalpa speciosa, and indeed all of the catalpas, can be readily grown from seed. In the late fall, beans containing the seeds turn from green to greenish-purple and weather into a light brown. The beans, when ready to harvest, must have a hollow sound when tapped. The ripe beans can also be judged by touch; they should have the feel of a rough, dry, well-cured Cuban cigar. On the tree, the largest beans will have the finest seed and these, in turn, should be collected from the midsection of the pod.

After harvest, the bean pods are stored in an open cardboard box, one layer thick, at 60–70°F (15–20°C), in diminished light, for up to three months. The dry seeds can be further stored at these temperatures in brown paper bags for up to two years.

After the drying period, the beans will begin to split open. A small full twist will open them completely. Inside, the seeds are long and flat with a tiny embryonic eye. Looking further, one sees that the top-quality seeds are found inside the bean semi-attached by dry friction. These have to be carefully scooped out by hand without twisting the seed, which will damage the embryo. First an edge is teased and then the seed will release itself.

At this stage of development, when the seeds are extracted from the beans, they are extremely dry. They then have some very interesting aerodynamic flight properties of lift based on electric charge and hydrophobic recoil. Consequently stray seeds are found in some very enterprising places despite their size. This is the ingenious seed dispersal method inherent in the genetics of this tree.

Catalpa seeds are planted flat and ½ inch (1.2 cm) apart in

pots. They are covered with ½ inch (1.2 cm) of potting soil. The pots are watered with lukewarm water and kept damp but not wet. They are placed in full sunshine. The seeds germinate in a very uniform manner. They take about 21 days to sprout. Growth is epigeal and very uniform. Two bundled cotyledons emerge that twist into a mop-head of four, just like a four-leafed shamrock. At this stage it is important to dry down the pots a little because the young seedlings are susceptible to damp-off. This fungal organism, *Pythium debaryanum,* lives at the wet interface of the seedling and the soil. A light dusting of dry woodash will prevent this fungus from injuring the growing seedlings. As the seedlings mature into the fourth leaf stage, they have left the threat of this fungus behind.

After the first killing frosts, catalpa seeds can be sewn outside. They can be planted a scant ¼ inch (1.2 cm) deep, 2 inches (5 cm) apart in rows 12 inches (30 cm) apart. In the colder zones 3–4, this can be as late as the end of May. They can be planted outside in mid-May for zones 5–7. For the warmer zones 7–10, planting around the beginning of April is usually successful. However, in all zones, one layer of Reemay will give some frost protection in the event of unexpected temperature dips.

The young seedling trees can be allowed to grow along the first summer. They are easily transplanted with a good root ball. Spring transplanting is more successful in the colder zones, while the warmer zones have a choice of spring or fall for success.

Catalpas prefer a pH of 6.5–7.5. They are calcareous species and grow well on limestone. They enjoy a rich, well-drained soil with a small amount of clay content, but they will also grow very well on lean, mean, rocky soils, provided these soils have winter drainage. Planting holes should have about 10 quarts (9.5 l) of well-aged sheep or horse manure, 1 quart (.9 l) of rock phosphate or .5 quart (.5 l) of steamed Canadian bonemeal, together with .5 quart (.5 l) of dolomitic lime. This is well mixed with the soil before the tree is planted. Trees growing in acidic soils can be limed with dolomitic lime to benefit the health of the tree. This can be done once in every five years. An increasing amount of dolomitic lime beginning with .5 quart (.5 l) can be broadcast

Long beans, produced by all of the *Catalpa* species worldwide

Catalpa speciosa in full bloom in July

around the tree from the collar area, where the soil meets the trunk, to the drip line.

Catalpas are exceptionally healthy trees both in North America and in Europe. However, they appear to be sensitive to sulfur dioxide and will not do well near a pulping or paper operation, or any industrial process that produces this acid-producing chemical. There is also a heartwood fungus called *Trametes versicolor* that can disfigure the trunk. This is prevented from entry by using rubbing alcohol or a 2 percent household bleach and water solution (1 oz/qt, 20 ml/l) on both cuts and pruning tools as an antifungal agent.

MEDICINE

The North American catalpa species hold many medicinal secrets. Some have been lost with the decline of oral history of the aboriginal peoples.

The medicine of the catalpa is found in the bark, the leaves, the twigs, the seed pod casings, and the seeds themselves. The biochemicals responsible for the medicinal effects are alkaloids such as catalpinoside, sugars, resins, pigmentation products, and highly volatile oils, among other as yet undiscovered biochemical complexes.

The giant tropical leaves have been used as a bandage wrap to draw an infected area. The inside of the leaf was used for a period of fifteen minutes. This was sufficient time to draw poison or a foreign agent from the limb.

A bark decoction was used as a tea. This was taken for the treatment of lung complaints such as asthma and bronchitis by the Cherokee peoples.

A tisane from the seeds was used as an antispasmodic and sedative, again by the Cherokee peoples. This medicine was taken in very small amounts.

The dry pod casings were used by the aboriginal peoples as an effective vermifuge to rid an area of parasites. These dry pods were slipped in among clothing and other household items to keep them free of pests, especially in the fall and winter months.

The catalpas produce an unusual biochemical called catalpinoside. This chemical has an extraordinary stereochemistry. The spatial arrangement of each chemical group is a three-dimensional, ultraviolet guiding mirror, whose function must be navigation. This is seen in the flowers. This guiding system for the visiting insect world is complete with sets of these chemical traffic lights guiding the insect toward the pollen arena inside the petal hood, then toward the female parts to complete fertilization.

Catalpas are also a source of antifungal and antiparasitic biochemicals. These are produced as a resin in extrafloral nectaries in the current season's bark and leaf petioles in glandular tissue. This purple resin is produced in trees that are exposed to the most sun.

Resins are used by the insect world to make their homes watertight, airtight, and defendable. These resinous mixtures are known as *propolis*. These are the medicines of the insect world.

In some years *Catalpa speciosa* produces a powerful fragrance, which appears to depend on moisture levels. This is probably an effort of the tree, in response to increased moisture, to switch on its terpene engine to induce flying visitors to help with greater pollination. This fragrance produced in abundance is very healthy for the lungs and the heart.

Bean casings have been woven into small baskets. These baskets have a natural pesticidal function and were used by the Cherokee peoples.

ECOFUNCTION

Catalpa speciosa is now uncommon in North America, but it is one of a number of trees that will cope well with climatic stress, especially in suburban areas. The catalpa is basically a tropical tree that has, by adaptation, moved itself into all of the north temperate regions of the world. It can withstand the ravages of sun, drought, snow, and ice. It produces a good-quality shade during the heat of summer. Because of its flowers and fruit it extends nature's food basket northward. Seed-eating birds such as the cardinals, *Cardinalis cardinalis,* have learned to use their beaks to split open the seed pods while they are still hanging on the tree. The cardinals release the seeds on to the ground and then eat them.

Even the young seedlings have learned to cope with climate change. The seedling, if it finds itself in a position of high solar radiation, will switch on a temporary sun screen chemical, going from green to purple. This happens in exactly the same way as in the seaweed genus *Rhodophyta* and the edible red seaweed species *Porphora gardneri.* This trick of survival chemistry might be just one of many of the catalpa species.

The huge, tropical leaves hang from an interesting stalk or petiole. The petiole is maroon on top and green underneath. These petioles serve as extrafloral nectaries that produce a purple overflow of a sweet resinous chemical that is much visited by flying insects. These nectaries occur at the axils of primary veins as they meet at junctions on the undersurfaces of leaves. They are more visible in September. These types of glands or nectaries make sources of food available to beneficial, nonflying insects also.

BIOPLAN

Catalpa speciosa is a feeding tree of ecological importance. It should be bioplanned into city and suburban forests as well as into the southern forests to increase biodiversity. The pattern of migration changes with global temperature fluctuations. This is presently seen in European bird populations. Bird habitat is changing with an average temperature change of 3.6°F (2°C) per one hundred years. Catalpas will keep pace with the needs of a moving bird and beneficial insect population.

Catalpa resin has an important function in the insect world. Insects collect an array of naturally occurring chemicals for their use domestically. Quite often these chemicals have finely honed

Digitalis purpurea 'Alba', white foxglove, an excellent garden companion to *Catalpa speciosa*

medical fungicidal and bactericidal properties as well as strongly hydrophobic properties. These chemicals are just as important to insects as the availability of nesting material is to a bird. Insects such as bumblebees and wasps use these chemicals in the "chewing gum mix," propolis, to seal the insides of their nest, making them waterproof. Insects also use them to control ventilation, which in turn controls nest temperature, which has to be regulated very accurately for brood rearing.

During the summer months when temperatures become elevated with increases in humidity, this propolis melts a little and puffs fungicidal and bactericidal aerosols throughout the nest, protecting inhabitants from infestations and disease. Resin-producing trees such as the *Catalpa speciosa* are an important part of an urban forest bioplan.

The catalpas are feeding trees for the insect and butterfly world. It is, however, the timing of bloom that is most important. At a time when nectar and pollen is drying up, the catalpa opens up its flowers, providing a source of protein-rich pollen and sugar to feed the summer replacement broods as the season progresses. The extrafloral nectaries pump sugar until the fall for feeding. This happens later in Europe for the same trees, extending into August, making this tree an important general store in European gardens also at a time of protein scarcity.

Catalpa seeds are eaten by many of the seed-eating birds. They are also collected for winter food stores by squirrels and the smaller mammals.

DESIGN

Catalpa speciosa is the tallest of the global catalpas; it can soar to 120 feet (37 m) in perfect growing conditions. It is a tree that is beautiful in all seasons for the gardener. In a northern garden, it will be the last tree to venture into spring foliage; this can happen as late as the beginning of June for zones 3–4. In colder gardens it stays a dainty midsized tree, a perfect complement to a two-story house in an urban setting.

When young, the catalpa has a rounded moplike appearance. This elongates with age into a curvaceous pyramid form. The branches become long and horizontal. These, when bound with swags of beans, knit the winter scene in a garden into one of curtained beauty. The lack of apical dominance creates an angular growth of twigs that switches the tree into a green cascade in the early summer, remarkable because of the wall of large, heart-shaped leaves. These soft green leaves turn to a full yellow in the fall.

Huge clusters of large, foxglove-shaped flowers adorn the trees in July. Each cluster has up to three different kinds of flowers that combine together to form a startling, white bouquet. The flower clusters hang on the tree for about two weeks. Inside the corolla of each individual flower the reticulation itself is beautiful.

In cooler, shady places under a *Catalpa speciosa, Digitalis purpurea* 'Alba', or white foxglove, can be planted. These biennials will seed themselves over time and will cope with dry shade quite well during the summer months. For sunnier sites, groupings of the catalpa with the hardy, native *Baptisia leucantha,* white or prairie false indigo, of the Leguminosae family give an easily maintained area in a garden where the underplanting does not need to be replanted and divided for about 20 years. The *Baptisia leucantha* has a flower cluster not unlike the catalpa flower, but without the markings and frills. In a small white garden, *Catalpa speciosa* could be used as an altar-form around which other white species could be planted, from annuals such as sweet alyssum, *Lobularia maritima;* white heritage roses, *Rosa spinossisima;* to white fragrant phlox, *Phlox caroliniana* 'Miss Lingard'; and white clematis, *Clematis virginiana.*

Catalpa speciosa can also be used to great advantage as an allée toward or away from a house or building. Over time they form a catalpa tunnel not unlike the rhododendron tunnels of the Edwardian gardens of England and Ireland of the last century.

For the smaller garden, *Catalpa fargesii,* zones 4–10, is becoming very popular in Europe, where it was first introduced from

Catalpa Species and Cultivars of Merit

SCIENTIFIC NAME	COMMON NAME	ZONES
Catalpa bignonioides	Indian bean tree	5–9
C. b. 'Aurea'	Golden bean tree	5–9
C. b. 'Koehnel'	Golden edged Indian bean tree	5–9
C. b. 'Nana'	Dwarf Indian bean tree	5–9
C. b. 'Variegata'	Variegated Indian bean tree	5–9
C. fargesii	Chinese catalpa	5–9
C. speciosa	Indian bean tree, orchid tree	4–9
C. s. 'Carrigliath'	Indian bean tree, hardy orchid tree	3–9

Baptisia leucantha, white or prairie false indigo, with an open flower not unlike the *Catalpa speciosa* in bloom

China in 1901. It won the award of merit in 1973. This tree is like a smaller version of *C. speciosa.* The flowers, instead of being white, are lilac pink with red-brown and yellow markings.

The southern native of North America, *C. bignonioides,* is a favorite, also, of European gardens. There is an outstanding golden-leafed cultivar, *C. b.* 'Aurea' and a strangely regimented form *C. b.* 'Koehnel', with leaves that have a broad outside margin of yellow. There is also a variety with creamy yellow leaves, *C. b.* 'Variegata'. As well, there is the beautifully formed, top-grafted, whimsical, umbrella shape of *C. b.* 'Nana' on a standard *C. bignonioides* to give a circular canopy. This graft experiment came out of France in 1850 and has delighted gardeners of small gardens in zones 5–10 ever since. This 'Nana' cultivar is particularly beautiful in a tiny garden near a small water feature.

Crataegus or hawthorn, species of which occur all over the north temperate regions of the world

Crataegus

HAWTHORN

Rosaceae — Zones 2–10

THE GLOBAL GARDEN

The hawthorn or *Crataegus* species occur all over the north temperate regions of the world. Luckily a great number of them have survived on the eastern face of North America from Florida up into the colder growing zones of Canada. Botanists think that there are about one hundred species of *Crataegus* native to North America. These trees have interbred over time, giving rise to a botanical stew. Genes from Ireland, England, and Siberia have popped into this stew, producing the aroma of a thousand hybrids. Many of the more important North American hawthorn species are held as collections in some of the older gardens of Ireland. To this day, however, no one in North America has sifted through this melee to compile a living arboreal collection of the hawthorn or *Crataegus* species for this continent.

The hawthorns range from rather squat, multistemmed, flat-topped shrubs to quality trees with a fine form. To gardeners, they are glorious, early-spring-flowering trees and shrubs bracing themselves to produce, through a cloud of thorns, an annual crop of applelike fruit. In the gulf states the fruit is known as the may haw from *Crataegus* species such as *C. aestivalis*, the shining hawthorn, and *C. opaca*, the apple haw. This is because these species slip into flower before May. In the remainder of the United States, the fruit of a hawthorn is known simply as a haw. In Canada these species are known as hawthorns.

The early English settlers are probably responsible for this great divide between Canada and the United States on the question of hawthorn, because these sailing refugees carried the middle English word *hawe* with them. The more northerly part of the continent went with the translation of the old Gaelic word *Sceach*, used so frequently in their starving past, meaning "thorny," "prickly," and "quarrelsome." So the Scottish and Irish immigrants used hawthorn for naming the fruit.

Historically, hawthorns, *Crataegus*, make up part of the patina of the English countryside. The hawthorns were, with the oak, *Quercus*, the ash, *Fraxinus*, and the birch, *Betula*, the pioneer trees that broke the way for the secondary growth after the ancient wildwood was felled in Mesolithic England around seven thousand years ago. In particular, the woodland hawthorn, *C. oxyacantha* (syn. *C. laevigata*) was and still is part of the ancient woodland and medieval hedges of England. The hawthorn can be used to date these hedges, many of which are more than five hundred years old. These are estimated in total to have been about 200,000 miles (321,860 km) long. Although some are truly ancient, most were created by the great Enclosures Acts of the eighteenth and nineteenth centuries.

Some mixing of tree species in hawthorn hedges date from Georgian times. But in Queen Victoria's reign, only hawthorn itself was used to make plashed hedges. These wonderful hedges are still stylishly laid to exhibition standards in such counties as Shropshire. Stakes and living ethers of hawthorn are woven to look like green basketry enclosing the fields.

In North America, the many hawthorn species found their place after the scouring of the last ice age. Hawthorns here too nursed the ancient wildwood back into being. The hawthorn, with its modified root growth that was tough and drought-resistant, was glad to colonize the miles of moraines and the glacial rock debris. They flourished on these barren places and hybridized in great glee because, genetically, marginal mothers make tough offspring. The great numbers of hawthorn hybrids are a statement to their colonizing success.

Curiously, the hawthorns of Europe and North America were

considered to be magical. The Druids imparted such strong warnings to the ancient world that in Ireland today many a child with a sprig of hawthorn bloom coming into the house is quickly told to go outside with it. To bring hawthorn into a house is to bring bad luck for an entire year. The aboriginal peoples of North America also had a belief that the hawthorn harbored bad luck. It would appear that the facts of fable sometimes hold a kernel of truth. Of the quarter of a million vascular plant species left on this planet, the hawthorn affects one little length of artery in the heart of man that could honestly be said to be the Achilles heel of mankind. So, yes, there is magic in the hawthorn.

There is great diversity in the North American hawthorns. A handsome 40-foot (12 m) tree grows around eastern Texas called the blue haw, *C. brachyacantha.* Its sibling is the apple hawthorn *C. aestivalis.* This tree also grows well in southern Ireland, producing a bigger, late fall crop of juicy, sweet-sour red apples than it does in its native Carolina. Probably the most common hawthorn, is the hog apple, *C. crus-galli,* or cockspur hawthorn. This species with dull red fruits is indeed beloved by porcine procurers. It is one of the oldest trees in North American hedgerows, tipping well over a hundred years. In many cases the thorns are useful barriers to wandering farm animals. Newfoundland sports the occasional, true, golden-fruited hawthorn, *C. chrysocarpa,* among the more common, dark red fruited kind. These straggle west to Colorado and can be found as far afield as Virginia, Missouri, and South Dakota. There is also a black-fruited hawthorn called the black hawthorn, *C. douglasii,* found in British Columbia, Alberta, and Saskatchewan. There is even a green-fruited hawthorn, *C. viridis,* lurking in the moist valleys of northern Florida that was named by the master himself, Carl Linnaeus, in 1753, four years before he rose to the Swedish nobility as Carl von Linné.

ORGANIC CARE

Pome is the technical term applied to a hawthorn fruit. The hawthorn is, like the apple, a member of the rose family. The pomes can occur singly on the branch or in large clusters. The fruit consists of an apple-type skin underneath which is a soft, sweetish flesh. At the core of the pome, there are one to four hard, light brown seeds called nutlets. In the south the fruit is picked in the early fall when it begins to soften by touch. In the north the fruit is picked after the first killing frost. This temperature drop both softens the fruit and sweetens it at the same time. The nutlets are separated from the fruit. Each nutlet will produce a hawthorn tree in time.

All hawthorn nutlets contain an embryo inside a grooved, hard shell that is not quite ready to germinate. A dormancy period is required during which the nutlets are stratified. All the nutlets of the southern, the northern, and the two European species, the woodland hawthorn, *C. oxyacantha,* and the common hawthorn, *C. monogyna,* require 30 days of stratification at 41°F (5°C) in a 50 percent damp sand-soil mixture in a refrigerator. These should be put into a sealed plastic bag, and as the earlier few seeds begin to break dormancy and to produce pure white radicles, they can be pricked from the soil mix and potted up immediately. If these "early birds" are left unnoticed, both they and the other nutlets will be spoiled by a number of pathogenic fungi.

Hawthorns all over the world exercise a tight, hormonal grip on the regulation of their germination. This discipline is based on long-term weather forecasts by means of enzymes. Consequently, the fruit crop of some years may take up to two years to break dormancy. There is no way of knowing this beforehand. This growth phenomenon is the survival strategy of the hawthorn.

The nutlets are planted out in the fall for the southern growing zones. They are spring planted as soon as the ground warms up in the northern growing zones. They are spaced 12 inches (30 cm) apart and covered with 1/3 inch (.8 cm) of soil. This soil is gently tamped by hand. The young hawthorns are allowed to grow on into the fall. Throughout the summer months, whatever the sapling size, they will have made considerable root growth. This extensive, adventitious root mass must be transplanted the following spring into either large pots or their permanent sites.

Hawthorns like a sandy, rocky soil of pH 6.0–7.5. The soil must be well drained in both summer and winter. Long periods of flooding appears to damage the hawthorn. The quality of the soil should be lean and mean, but the addition of 2 quarts (1.9 l) calcium in the form of ground dolomitic limestone is much to their liking, as it is an aid to the future production of fruit, for which this mineral is required. This soil conditioner can be added at planting time or can be broadcast around the tree at any time thereafter.

The problems of the hawthorn are weather dependent. Sometimes they are affected by fungi, bacteria, or insects. During hot, humid summers, the hawthorn rusts, *Gymnosporangium clavipes* and *G. globosum,* can be seen to disfigure leaves and fruit late in the season. The following year they are gone. The same can be said for the leaf blight, *Diplocarpon mespili.* These also disappear the following spring. Hawthorn can benefit from a lime-sulfur dormant oil spray in the early spring. The ground around the hawthorn can also be lightly powdered with dry woodash. This can be extended up the trunk to kill off ascospores and prevent them from reproducing.

The two European hawthorns, the woodland hawthorn, *C. oxyacantha,* and the common hawthorn, *C. monogyna,* both respond to the higher solar conditions of North America with more aggressive growth. This, in turn, makes them more prone to a bacterial killer called fire blight, *Erwinia amylovora.* The pruning of these trees should be kept to a minimum. When and if they are being pruned, the cutting tools should be extremely sharp because the wood of these two hawthorns, indeed like the majority of these species, is rock hard. A sterile technique using a 2 percent household chlorine bleach and water solution (1 oz/qt, 20 ml/l) for tool and cut disinfection should be practiced while cutting. Hands, too, should be clean, because this bacterium can travel on hands or clothing.

Crataegus in bloom. When in flower, many *Crataegus* species produce a strong fragrance that is healthy for the heart. This fragrance is trapped by the membranes in the nose.

The fruit of the *Crataegus crus-galli,* the cockspur hawthorn, ready to eat after the first frost when the sugar content increases. It is an ideal trail snack for man and creature alike.

MEDICINE

Hawthorn, *Crataegus,* has been used as a medicine for a very long time. It was well known to the ancient Greek herbalists. It was used in Ayurvedic medicine. It was used in Europe, northern Africa and western Asia. It was also known to the Druids and the aboriginal peoples of North America. Hawthorn, *Crataegus,* has also been associated throughout the ages and all over the world with the practice of magic.

C. calpodendron (syn. *C. tomentosa*), *C. punctata,* and *C. submollis* were commonly used by the aboriginal peoples of North America. The woodland hawthorn, *C. oxyacantha,* and the common hawthorn, *C. monogyna,* were used in Ireland, England, and European countries for medicine. Principally their use was as a hypotensive agent.

In North America, the bark, root bark, leaves, fresh and dried fruit, together with the nutlet seeds were used as a source of medicine.

In Europe, the fresh and dried fruit was used. The constituents of *C. oxyacantha,* the woodland hawthorn, are available as commercial extracts of fruit, flowers, and leaves. These extracts are known as Curtacrat, Crataegus-Kreussler, or Esbericard (all three are trademarked names). These extracts are used in modern medicine as cardiotonics and also as effective coronary vasodilators.

The constituents of *C. oxyacantha,* the woodland hawthorn, include triterpene acids called oleanolic, ursolic, and crataegolic acids, as well as purines and anthocyanin type pigments. There are also flavone derivatives called pelargonin and quercitrin. Choline, acetylcholine, trimethylamine, and chlorogenic acid also occur. There is caffeic acid and vitamin C as ascorbic acid. A substance called temporarily RN 30/9 has been isolated that acts as a stimulating growth hormone for all caterpillars. In the North American hawthorn, the triterpene content could be expected to be greater. The anthocyanins will vary from species to species. The flavones will be greater in the southern species, the choline compounds higher in the north.

All of the species of hawthorn, *Crataegus,* were used by the aboriginal peoples as a food either fresh, dried, or sweetened with maple syrup. They were also used by the farming community, the ripe, frosted, sweet berries eaten out of hand as a trail snack.

The Meskwaki aboriginal peoples used the unripe fruit of *C. calpodendron,* the pear hawthorn, as a diuretic in the treatment of bladder ailments. The fruit clusters, which grow in an upright position on the branch, were picked just before they turned an orange-red color in summer. These pear-shaped fruits had a thin sweet juicy pulp. This pulp was eaten raw as a medicine, the seeds spat out.

The Chippewa used a decoction of roots of possibly shining hawthorn, *C. aestivalis,* in the treatment of back pain. A handful of roots were boiled thoroughly in a quart (1 l) of well water. A quart of the decoction was taken on a daily basis.

The Mohawk aboriginal peoples used the dotted hawthorn, *C. punctata,* together with the garland tree, *Malus coronaria,* as a hypotensive agent to stop menstrual flow. A handful of young shoots and bark was taken from *M. coronaria*. This was added to bark chips taken from *C. punctata*. The chips were taken using an upward cut. These were boiled in .5 gallon (1.9 l) of well water. The decoction was further reduced to 1 quart (.9 l). This water was taken as needed. *M. coronaria,* the garland tree, is the wild apple, native to eastern America. This species is getting exceedingly rare in the wild.

The Cayuga aboriginal peoples used *C. submollis* together with the common weed called pig weed, *Amaranthus retroflexus,* in a magical antidote to being lovelorn.

The habit of smoking is as old as North America itself. Many of the hawthorns, *Crataegus,* were used as a stimulating substitute for tobacco. The mature leaves were collected and dried. These were coiled and rolled into a cigarette. Occasionally these cigarettes were also seasoned with the juices of local fruits and dried again. A sweet, stimulating smoke was enjoyed by many of the North American aboriginal peoples during the long winters.

A cup of coffee, too, was available to the aboriginal peoples. All of the North American hawthorn, *Crataegus,* species have nutlets that are high in caffeine. The seeds were collected; sometimes they were roasted. They were ground and an infusion was made with a coffee flavor that was just as stimulating as modern-day coffee.

The woodland hawthorn, *C. oxyacantha,* and others seem to have an effect on angiotensin II, which is a potent blood vessel constrictor. It also seems to have a stabilizing effect on the heart muscle itself and be of aid to paroxysmal tachycardial arrhythmias. Thus, the biochemical complex of hawthorn, *Crataegus,* appears to have a beneficial effect on stage II congestive heart failure by its hypotensive action.

All in all, the stage is open for the medical testing of approximately one thousand hybrid *Crataegus* species as new hypotensive agents with a specific target of improving the flow of the left ascending coronary artery of the human heart. It is the clogging of this little strip of artery that requires a surgical procedure known as bypass surgery to repair damage.

ECOFUNCTION

The hawthorn, *Crataegus,* makes way for the forest in a most unusual way. The collar region of the hawthorn, that is, the area where the soil meets the trunk of the tree, has a damp "drop" pocket. This opens and closes depending on the time of the year and the moisture content of the soil. The "drop" pocket captures maple seeds.

The maple needs a nurse crop to provide shade to young sapling trees. The hawthorn provides the correct amount of shade. The maple can compete with the hawthorn. As the tree grows the hawthorn dies. But in death it protects the maple even further because the dead thorns are even more treacherous than live ones. So the young maple is protected from girdling. Underground, maple and hawthorn share a unique type of mycorrhizal growth for root meristem feeding. And perhaps the death of one enhances the feeding fungal sheath of the other.

All of the hawthorn, *Crataegus,* leaves and fruit produce a number of biochemicals that assist in the migration of mammals, birds, and, more uniquely, butterflies. Choline and acetylcholine found in high concentrations help in the formation of a high-energy compound called adenosine triphosphate, ATP, which is the fission fuel for providing the energy or power to migrate. The larger animals require this, as do such songbirds as the thrashers and mockingbirds, catbirds, bluebirds, shrikes, grosbeaks, cardinals, blue jays, and mourning doves. Then the hawthorn doubles as an escape and very safe nesting site. This is because the design of the thorns on the branches is zigzag so that rapid climbing by anything other than small mammals is out of the question.

The mature leaves of the woodland hawthorn, *C. oxyacantha,* produce a hormone that benefits the butterfly population. The hormone is a substance called RN 30/9. This is a biochemical that boosts and enhances the growth of the young caterpillar, making it a stronger butterfly to both fly in migration and enhance its ability for survival. It could be reasonably expected that the southern state hawthorns and hawthorn species that

Crataegus crus-galli, the cockspur hawthorn, has long been associated with the extraordinary songbird, the loggerhead shrike or "butcher bird," that impales its prey on the long thorns.

flank great bodies of water would also produce other hormonal biochemicals as boosters to butterflies for long-distance travel. Hawthorn plays host to some wonderful swallowtail butterflies. One is called the white admiral, *Basilarchia arthemis,* and its flight companion is the red-spotted purple swallowtail, *B. astyanax*. This swallowtail tries hard to mimic another very poisonous butterfly called the pipevine swallowtail, *Battus philener*. The colors are reproduced perfectly, but the patterning is a little different. The hawthorn also plays feeding host to the striped hairstreak, *Satyrium liparops,* and to the more rare northern hairstreak, *Euristrymon ontario*.

All species of hawthorn produce an abundance of flowers in the spring. They look much like apple blossoms, to which they are closely related. Their fragrance is not sweet like the apple, but is strong and vibrant. The majority of the hawthorn blossoms are of a single, chalice-shaped form with the male anthers and the female receptacle at the center. The petal color can range from white to pink to scarlet. The overall morphology of the flower makes it perfect for insects to maneuver and feed for either pollen or nectar.

The parabolic form of the flower is ideal for emitting the chemical message of fragrance, which, in this instance, is composed of three volatile terpenes. Honeybees, bees, wasps, and other flying insects receive this chemical message and come to pump up. But should the weather become cloudy, the chemistry of smell is switched off. The male anthers then flop over the nectar source. The insects are cut off from feeding. This strategy manages a very effective pollination and cross-pollination campaign for the hawthorn.

BIOPLAN

In North America the hawthorn, *Crataegus,* should be bioplanned into hospitals, clinics, or senior citizen and retirement housing complexes. The downy hawthorns, *C. mollis* and *C. submollis;* the Biltmore hawthorn, *C. intricata;* and the pear hawthorn, C. *calpodendron;* could be used. The flowers of these species produce a strong fragrance that is higher in the volatile triterpene complexes because of their xerophytic characteristics. These chemicals are absorbed by the mucosal membranes of the nasal passages and become a "passing-by" cardiotonic for cardiac recovery patients. These hawthorns could be planted in multiples around a sitting area positioned due south. The trees should be spaced 10 feet (3 m) from the seating arrangement. The hospital or clinic wall should be to the north to reflect and amplify the air.

The majority of the North American hawthorns are important environmental species. They act in erosion control of sandy soils. They grow well in maritime areas and act as wind buffers.

They are useful feeding and nesting trees to bring songbirds into the urban forest. If planted in public parks, they reduce the need for pesticide use because they harbor predatory species and therefore maintain a balance for predation.

The farming community will be among the first to feel the strain of global warming. Crop damage will increase due to wind, wind bursts, and periods of drought and wet associated with warming trends. These will all help viral, bacterial, and fungal pathogens to multiply. New regimes of noxious insects also migrate with climate change. Bioplanning hawthorn species into hedging for smaller fields, as wildlife feeding boundaries, acts as a reservoir for predation. The North American hawthorn species are high on the list for multiplying beneficial predation by songbirds and insects.

The hawthorn, *Crataegus,* species make excellent bee pasture. The hawthorn pollen is cream colored and is like a miniature of the nutlet seed. This pollen is much sought after by honeybees, as is the nectar. The honey produced from hawthorn is of a very good keeping quality and is not unlike, in flavor, the honeys from the heath or Ericaceae family.

The sweet pulp of the hawthorn pome could supply a new kind of food for athletes if used as a fruit leather or mixed with other foods. The biochemistry of the pulp would improve athletic performance naturally.

The horticultural world could gain much from the select breeding of the succulent hawthorn, C. *succulenta,* and the beautiful hawthorn, C. *pulcherrima,* just to name a few. These could be bred for a new fruit and horticultural flowering specimen, respectively.

In the past, all over North America the hawthorn fruit was used to make a high-quality jelly that was rich in natural pectin. The pome of the hawthorn also makes a delectable liqueur.

DESIGN

There are many stars in the galaxy of *Crataegus* species from which to choose. A species with the most attractive flowers is the beautiful hawthorn, C. *pulcherrima.* It has large ¾-inch (1.9 cm) white flowers with purple stamens. The species with the finest autumnal coloring of scarlet and orange is the Washington hawthorn, C. *phaenopyrum.* An outstanding cultivar for a patterned bark is the green hawthorn, C. *viridis* 'Winter King'. The exfoliating, silver-gray bark reveals an orange-brown inner bark that is quite handsome. This is topped off in the early winter months by a swag of very large, orange-red fruit.

Probably the most useful, hardy species for north temperate gardens is the cockspur hawthorn, C. *crus-galli.* This tree can be trained to have a horizontal plane of white bloom in the spring. There is a thornless cultivar C. *c.-g.* 'Inermis' of the cockspur hawthorn and a hybrid, C. *c.-g.* × C. *phaenopyrum,* called 'Vaughn'.

The cockspur hawthorn and its cultivars should not be planted as companions to the domestic apple, the *Malus* species, because this hawthorn can co-host an apple-cedar rust fungus.

For the warmer gardens of zones 6–10, the blueberry hawthorn, C. *brachyacantha,* is a remarkable tree massed with blue berries carrying a fine dusting of bloom following a crop of white to orange blooms.

The two English hawthorns, C. *oxyacantha* and C. *monogyna,* are more commonly found in the older garden designs of

The white flowers and red berries of *Smilacina racemosa,* false Solomon's seal. Both add a season of attention to all of the *Crataegus* tree species.

North America. Despite their great beauty and vigor they are not as hardy as the native species, being quite often damaged by severe winters in zone 5. However, they have some very fine cultivars. *C. m.* 'Compacta' is ideal for a small garden in zones 5–10. The woodland hawthorn, *C. oxyacantha,* has a wonderful double flowering form, *C. o.* 'Paul's Scarlet', which has long been a favorite of gardeners on both sides of the Atlantic. This tree arose as a natural sport in a garden in Hertfordshire in England in 1858 and comes true from seed. It is a magnificent tree in full bloom, the whole form shrouded in red. There is a double white *C. o.* 'Plena' (syn. *C. laevigata,* 'Plena') and a double pink *C. o.* 'Rosea' (syn. *C. l.* 'Rosea') and a cultivar with bright yellow berries *C. o.* 'Lutea' (syn. *C. l.* 'Lutea'). There are also weeping, fastigiate, and tortuosa forms. At Morden, Manitoba, *C. oxyacantha* 'Paul's Scarlet' was crossed with *C. succulenta.* Two very hardy prairie hybrids were produced, one a double white called 'Snowbird' and another double white turning shell pink called 'Toba'.

Hawthorns, *Crataegus,* add a fine skeletal form to the fall and winter garden with a touch of color from the resident hanging fruit. These pinpoints of red can be picked up and massed with the bunchberry, *Cornus canadensis,* with its scarlet heart of fruit at ground level. Or airborne files of drooping berries can be used with false Solomon's seal, *Smilacina racemosa,* to add fragments of red. The shademaster ginsengs can also be used, the native *Panax quinquefolius* and *P. trifolius,* whose clutches of red berries sing a medicinal song to the haws nearby.

Crataegus Species and Cultivars of Merit

SCIENTIFIC NAME	COMMON NAME	ZONES
Crataegus aestivalis	Apple haw	5–10
C. brachyacantha	Blueberry hawthorn	5–10
C. calpodendron	Pear haw	5–10
C. chrysocarpa	Roundleaf hawthorn	2–10
C. crus-galli	Cockspur hawthorn	3–10
C. c.-g. × *C. phaenopyrum* 'Vaughn'	Vaughn Washington hawthorn	3–10
C. c.-g 'Inermis'	Thornless cockspur hawthorn	3–10
C. douglasii	Black hawthorn	3–10
C. intricata	Biltmore hawthorn	4–10
C. mollis	Downy hawthorn	3–10
C. monogyna	White thorn	5–10
C. oxyacantha (syn. *C. laevigata*)	Woodland hawthorn	5–10
C. o. 'Lutea'	Yellow English hawthorn	5–10
C. o. 'Paul's Scarlet'	Paul's scarlet English hawthorn	5–10
C. o. 'Paul's Scarlet' × *C. succulenta*	Toba hawthorn	3–10
C. o. 'Paul's Scarlet' × *C. succulenta* 'Plenum'	Snowbird hawthorn	3–10
C. o. 'Plena'	Double white English hawthorn	5–10
C. o. 'Rosea'	Rose English hawthorn	5–10
C. phaenopyrum	Washington hawthorn	4–10
C. pulcherrima	Beautiful hawthorn	7–10
C. viridis	Green hawthorn	5–10
C. v. 'Winter King'	Winter King green hawthorn	5–10

Hawthorn berries can be used with false Solomon's seal, *Smilacina racemosa*, to add fragments of red to the fall garden.

A giant American white ash, *Fraxinus americana*, that was used as a survey marker for the Rideau Canal system in the 1830s. It resides with its Christian companion, St. James the Impossible, in the little village of Burritt's Rapids, Ontario.

Fraxinus

ASH

Oleaceae

Zones 3–10

THE GLOBAL GARDEN

The ash is considered to be the primordial tree from which the human race was once plucked. This idea seems to exist across the world. In an Algonquian legend this happened with their fabled warrior, Gooscap. He shot an arrow that was made of ash into an ash tree to spill out the first human beings. This is somewhat repeated in northern Europe by an old Norse saga in which the mythical, primordial ash tree was called Yggdrasill. This tree had the souls of the unborn still in its limbs. Then there is the story of the self-crucifixion of the German god Woton on the primordial ash tree. This self-sacrifice was in exchange for runes. Runes were the ancient symbols of the early alphabet of Western literacy. Even the Egyptians seemed to be familiar with ash trees, for they are to be found tabulated carefully in housekeeping hieroglyphic records.

Scholars now think that the manna mentioned so often in the Bible is in actual fact an ash tree. Manna was miraculously dropped from the sky to the Israelis as they fled, starving, through the wilderness on their way to the Holy Land. There is an ash tree native to the Middle East called *Fraxinus ornus,* the flowering ash or the manna ash. This little tree, if cut, will produce a sap exudate somewhat similar to *Acer saccharum,* the sugar maple. This sap dries rapidly into a crystalline mixture of mostly a sugar called mannitol and its polymers. There is probably not a better energy source than this sugar for somebody on the run. The sap from *F. ornus* is still in use as a food in parts of southern Italy, as it was around the time of Christ.

While the black ash, *F. nigra,* is a vital ceremonial tree to the aboriginal peoples of North America, the European ash, *F. excelsior,* was an important tree to medieval trade and commerce. In England the growth of the ash was modified by a special technique called coppicing. Woods were coppiced on huge tracts of land. Some of these ancient woodlands remain much as they once were described in the Domesday Book of 1086. Hatfield Forest, now a property of the National Trust, is one of the most famous.

The early Greeks used *F. excelsior,* the European ash, in magical rainmaking rituals. This was because this ash tree, in Europe, received so many lightning strikes during a thunderstorm. The ancient Irish used wood from *F. excelsior* to make hurleys for their rapid-fire ball games of hurling for men and camogie for women. They called the wood *fuinseóg* and the implements made from it *fuinse.* This was clearly borrowed from their old Gallic allies, the French, the old French name for ash being *fraisnaie.* This is almost identical in spoken language to the old Gaelic form. It is repeated as *fresno,* in Spanish.

The ash species help to make the backbone of the north temperate forests. There are about sixty-five species of ash in the global garden. Four of these are common to eastern Canada and a further twelve native to the United States. In Canada, the ash are declining through logging and a universal, unidentified disease that is whittling away at the health of these trees both in number and in ability to survive. The giant specimens of black ash, *F. nigra,* which were used as plates for roof support in the old-style cedar log barns, are rarely to be found growing in the wild. However, the very cold-hardy black ash is still the most common ash native to Canada. Despite its prominence as an important tree of the ancient virgin forests of Canada, this tree receives little respect. Very little is known about the various sexual races of male and female trees existing on the continent. One of the fastest-growing deciduous trees is the white ash, *F. americana.* These trees grow in abundance around the Great Lakes and the St. Lawrence River valley and throughout Nova Scotia, New Brunswick, and Prince Edward Island. It is found all over the eastern United States as far south as northern Florida. The hardy red ash, *F. pennsylvanica,* is to be found on the riverbanks of the Saskatchewan prairies east into Nova Scotia and down into Texas and the north of Florida. A rare blue ash, *F. quadrangulata,* is surprisingly also native to Canada. It grows in a solid grouping around the Thames River in southern Ontario between London and Chatham. The blue ash is also found on Point Pele and its island. It also grows in West Virginia.

There are a number of ash species of interest in the United States, the Carolina ash, *F. caroliniana,* and the two-petal ash, *F. dipetala,* of California and Mexico. There is also the desert or velvet ash, *F. velutina,* which is a tough, drought-resistant species. The evergreen, heat-loving shamel ash, *F. uhdei,* of central Mexico and Hawaii is often planted on the streets of California.

Fragrance is found in many species of nonnative ash. The Chinese ash, *F. chinensis,* and the Marie's ash, *F. mariesii,* are both beautiful flowering ash trees, as is the manna ash, *F. ornus,* of southern Europe and Southeast Asia.

Historically ash has been used in the sports world from hurleys to lacrosse and hockey sticks to tennis rackets and polo mallets. It is also used in playgrounds where the wood is expected to survive the daily hammering of young children. It is used for chairs. It is bent for snowshoes and it is woven into seats. It has been brought to a high art form in the popcorn weaving of the ceremonial baskets of the aboriginal peoples. Many of these are now in museums. They are reminiscent of Gaelic wool work that is presently found on the Aran Islands off the coast of Ireland. The language of family sorrow and legends was knitted into the raised berry work of the sweaters much like the popcorn weaving of the aboriginal black ash baskets.

ORGANIC CARE

All ash trees can be grown from seed. The seeds are called samaras. They are singular and hang from the trees in fist-sized clusters that are clearly visible. Ash trees are easily transplanted with a minimum of transplanting shock and will grow happily in a great variety of habitats, from wet to dry and in sun or shade. Ash trees can be coppiced and can be pollarded.

European, Manchurian, and Chinese ash grow well in North America, and *Fraxinus excelsior,* the European ash, is often used in landscaping here. This tree grows with a denser upper crown than *F. americana* and is a better shade tree. In addition, the side branches reach out horizontally to the sunlight. If seed is to be collected from these foreign species, it can be done while the seeds or samaras are green. They must be allowed to ripen to a brown color in dry, sunny conditions. After the point of total color change, the samaras must be left for a minimum of 14 days in the same sunny dry conditions to allow the white embryo inside the samara to continue growing and to mature. This is because dormancy of these species is different from the dormancy conditions of the North American native ash species. The samaras have, at that time, reached their greatest germination potential.

All native North American ash species can be collected in the fall when they have turned brown. At this stage they will have a pure white embryo inside. Both native and nonnative species should be planted flat, about 20 seeds per 12 inches (30 cm). They should be covered with about ¼ inch (.5 cm) of light soil. Then 4 inches (10 cm) of dry crinkly leaves should be placed on the planting rows as surface heat sinks to help the samara's epigeal germination. In zones 3–5 these young seedlings can be protected from harsh winters by dropping a makeshift cold frame around them using anything at hand from bales of hay to blocks of wood topped by a spare pane of glass. This also allows the seedlings to grow on in the early fall and to harden off nicely before transplanting in the spring.

Because the North American continent has been trained by frequent, naturally occurring forest fires and because all of the ash species are damaged by fire, especially when young, the ash instituted various techniques for survival. The *F. americana,* the white ash; *F. caroliniana,* the Carolinian ash; *F. pennsylvanica,* the green ash; *F. uhdei,* the shamel ash; and *F. velutina,* the velvet ash; all have both male and female trees. The blue ash, *F. quadrangulata,* has both male and female parts in the same flower. Therefore each tree is potentially fertile. This is also the situation with many ashes like *F. dipetala,* the dipetal ash, in California. This diverse sexuality helps in species survival.

In the case of *F. nigra,* which is probably the most prolific ash on the continent, both male and female trees appear in some areas of North America. In other areas, trees have male and female organs in the same blossom. Therefore each tree is self-fertile. In either case, the male pollen may be sterile, or pollen may not be produced in sufficient quantity to pollinate, or the timing of pollen release is either too early or too late to affect pollination, leading to infertility.

From a gardener's or collector's point of view the female samara bearing trees must be located by July. The trees are watched for samara development weekly. A number of samaras are opened lengthwise to reveal a white embryo slip that should look like damp, shining porcelain, indicating that all of the samaras are viable seeds. This is true in all of the ash species.

Ash seeds with intact wings, or pericarps, can be stored over the winter months in glass jars held in a refrigerator at 41°F (5°C) until spring. They are then planted 20 seeds per 12 inches (30 cm) and covered with ¼ inch (.5 cm) of soil. In all cases the older an ash samara is, the longer its dormancy. Old ash seeds can be stored at 41°F (5°C) and spring planted. These seeds may take several years to germinate, so the propagation area should be well marked.

Ash trees like a well-drained, fertile soil preferably with an underlay of dolomitic lime and a pH of 6.0–7.5, or around neutral. Under these circumstances, the ash will grow to its full potential. However, ash will grow in damp areas, with standing water or on dry, barren, rocky sites. Ash can even be found in swamps, particularly *F. nigra,* the black ash.

Garden soil is amended with dolomitic lime, about 1 pint (.5 l) per hole, or dry woodash, 1 quart (.9 l) per hole. Two quarts (1.9 l) of good leaf mold or garden compost is added and mixed with the soil. A pint (.5 l) of steamed bonemeal or rock phosphate is added and mixed. The tree is set in place and a small depression is left to catch water. In dry areas the young tree can be mulched with stones.

Ash trees, like their cousins the olives, are rarely seriously attacked by insects. However, recently new insect pests from other countries are having devastating effects locally in Michigan. Hopefully, trees will develop some resistance over time or predators of the pests will begin to control these situations. But the ash since the 1940s have been suffering from a dieback. It seems to start with the lower branches. These, at the peak of their growth in midsummer, start to shed leaves, which turn yellow. This is followed immediately by the death of the branch or branches. The tree limps on in growth, having lost much of its vigor. The disease is certainly systemic, possibly involving a climatic cycle of drought and internal problems that are possibly soil-borne viruses. Some plant pathologists think that *Cytophoma pruinosa,* a cancer-causing fungus, is involved. It is possible that the ashes are responding to an increase in carbon dioxide and general acidification affecting mycorrhizal growth.

The black ash, *Fraxinus nigra,* the choice spot for a vireo nest in the forest. There is a stillness in the tree that helps to make this valuable songbird invisible while nesting.

MEDICINE

The ash species have been used in global medicines in three principal arenas, in kidney treatments, as a laxative, and as a vasopressor.

The deciduous holly, *Ilex verticillata,* winterberry, is often a marker of *Fraxinus nigra,* black ash, habitat.

The principal constituents of *Fraxinus* will vary with each species, and the growing location of the species will influence the content of carbohydrates with their isomeric forms. The ash contains several sugars. These range from mannitol, mannotriose, mannotetrose, and dextrose to glucose and glucuronic acid. It also contains mucilage, escin with its complex glycosides, esculetin, and esculin. There is also fraxetin with its glucoside fraxin.

The bark of both the North American, *F. americana,* and the European, *F. excelsior,* was used to treat hemorrhoids for centuries. Bark infusions of *F. americana* were used to treat skin sores and to cure itch by the Meskwaki.

As a laxative it was considered by the Iroquois to be mild and gentle and of use in convalescence. Fraxin is the principal ingredient. A bark bundle was selected of *F. americana* consisting of young shoots of the same diameter as the combined thumb and forefinger. The bark was scraped downward from these shoots and saved. It was boiled in 2 quarts (1.9 l) water reduced down to .5 quart (.5 l). This was drunk as a laxative to reduce cramps.

It was also used as a laxative for horses. Two double handfuls of bark bundles were boiled in 6 quarts (5.7 l) of water. With the bundles removed, the liquid was reduced to 1 quart (.9 l). This was added to the horse's daily rations.

The bark of *Fraxinus* was used as a hunting medicine for deer. It was also used for fishing and to stop being bitten by snakes. This medicine was considered to be good luck. In fact, there is a biochemical basis for this trick. The bark of *Fraxinus* species have escin, which is a potent biochemical complex used in the treatment of peripheral vascular disorders. This complex functions as a vasoconstrictor, slowing down the skin heat and hence the human odor that is volatile and is dependent on skin temperature fueled by the surface circulation. The creature being hunted cannot smell the human odor and therefore the "luck" of the hunter increases.

The white ash, *F. americana,* is also used in the treatment of snake bites. A "strangle-root" is located around the root area. This root will be looped above the ground. The bark is kicked off rapidly and applied directly as a poultice with some warm water to the area. The root decoction can also be drunk.

The high content of D-mannitol in the ash makes it a potent diuretic for increased renal function. This sugar is found in varying concentrations in all of the *Fraxinus* species. It is particularly high in *F. ornus;* the bark extract contains from 46 percent mannitol.

Both *F. nigra* and *F. americana* are used in a combination with other native species by the Seneca as an interesting means of inducing pregnancy. A handful of bark from the east side of butternut, *Juglans cinerea,* with the east-facing roots of leatherwood, *Dirca palustris,* curled dock, *Rumex crispus,* burdock, *Arctium tomentosum,* is mixed together with a handful of bark of *F. nigra* and *F. americana* and an entire plant of pipsissewa, *Chimaphila umbellata,* of the wintergreen family. This is boiled in 8 quarts (7.6 l) water down to half volume, 4 quarts (3.8 l). A tablespoon is taken before meal time, the rest being up to the aspiring parents!

ECOFUNCTION

There are a few species of ash that are oil-bearing trees. In China, one species is bled for its oil and waxes, which are used in candle making. All of the native north temperate species of ash, when split open, expose a wood that seems to have a fine film of oil on the surface of the wood itself, making it shine in full sunlight.

This is probably why ash is so easy to split and why ash is one of the few woods that will burn in the fresh, green state. This phenomenon is true of European and North American species. In Ireland, ash twigs were used to "lighten-up" a slow turf (peat) fire that had been smoldering due to dampness. This was a trick needed for open-fire bastable cooking in the past. It was adopted by the pioneer cooks in Canada, using white ash. The oily chemical characteristic of the wood makes it difficult to glue but excellent for baskets, some of which could hold water. It is the reason why many species of ash thrive in damp or wet habitats. It could also account for the fact that no observed higher order species of fungi are associated with the mycorrhizal root growth of the ash underground.

As the bark of the ash gets older, it produces long triangular spaces that are ideal for the growth of mosses and lichens. These species are always constant for the ash wherever it happens to be growing. The mosses and many species of lichens found growing on ash add considerably to the biodiversity of the forest in increased insect populations and are excellent indicators of air pollution.

The samaras of some ash species are fragrant, for example *F. quadrangulata,* the blue ash, and *Fraxinus nigra,* the black ash. This is a chemical calling card for feeding by the seed-eating songbirds such as the evening grosbeak and the waxwings.

The female flowers of ash are open in construction, being devoid of petals or sepals, and are visited by honeybees for nectar.

The ash plays host to many species of butterflies, the tiger swallowtail, *Pterourus glaucus,* and the two-tailed swallowtail, *P. multicaudatus.* It hosts the hickory hairstreak, *Satyrium caryaevorus,* and the wonderful Baltimore, *Euphydryas phaeton,* visitors of damp meadows and boggy areas.

The ash species have many microscopic forests on the apical bud structures and under the leaves that act as reservoirs of protection for the smaller insects of prey. These help to keep the harmony of nature in a healthy balance within the forest ecosystem.

The ash is a shade-giving tree that can withstand considerable sunlight. As atmospheric changes proceed with ozone depletion, there is an automatic increase of ultraviolet light. Excess ultraviolet light is not beneficial to most higher life forms. But the ash species have their own light filtration process in the form of a chemical called esculetin (6,7-dihydroxy-2*H*-1-benzopyran-2-one). It is found throughout the tree. The form of the double ring structure is such that it traps ultraviolet light by electron resonance. This makes the ash a very important tree for shade.

The blue ash, *F. quadrangulata,* produces a blue dye. This is found underneath the bark like a continuing epidermis for the tree. The dye was an important color for the aboriginal peoples' handiwork. It is released into water by stirring some freshly broken blue ash twigs.

The black ash, *F. nigra,* or *Ehsa,* as it is called by the Mohawk nation, is important for basket making. A 30-foot (9 m) straight log is selected. This is cut and pounded longitudinally. The winter and summer woods separate. Strips of varying thickness are cut and then may be further split. These are dampened and woven with different colored, naturally dyed strips into complex baskets and serve symbolically and practically in the ceremonies and rituals of life. Thus the ash becomes a fundamental element in the richness of aboriginal culture as a part of the fabric of the religious iconography. Basket making, too, is passed down the generations. No part of the *Ehsa* is lost, for the pith is burned as wood. These wood ashes are carefully collected. This dry ash is high in potash. It is one of the essential ingredients in the preparation of their corn dish known as hominy.

Wild coffee, *Triosteum aurantiacum.* With red flowers and orange berries, it is the native companion to the ash, *Fraxinus,* in the wildwoods of North America.

The samaras of *Fraxinus americana,* American white ash

BIOPLAN

The ash are excellent colonizers. They are some of the first hardwoods along with basswood, *Tilia americana,* to begin the long trek of reestablishing hardwood forests. The shade from the pinate ash is dappled and moving, making an ideal nursery for all of the acorn-bearing trees.

Ash should be planted into farmers' hedgerows. This tree will help to reduce the damage inflicted on crops by ultraviolet spectra. Pollen tubes, produced by pollen in the process of fertilization of ova, are being damaged by direct exposure to the ultraviolet spectrum. This in turn affects crop productivity, which can be somewhat alleviated by the old-fashioned ash hedge. Farm animals, especially short-haired breeds, are becoming prone to cancerous skin melanomas. Again, shading by ash trees will help reduce this potentially fatal condition.

Female cultivars of *F. americana,* the white ash; *F. pennsylvanica,* the green ash; *F. nigra,* the black ash; and *F. caroliniana,* the Carolina ash; can be used for school yards, day-care centers, and kindergarten schools. These female cultivars cannot produce wind-borne pollen, which causes hay fever in some people. The trees themselves would act as excellent ultraviolet shade filters for young, developing babies and children who are very susceptible to this fraction of the light spectrum. Zoos and public parks would greatly benefit from the shade of the ash.

Extensive trials on the black ash, *F. nigra,* are under way to replenish the diminishing supply of this species at the Akwasasne Reservation.

Ash is an excellent environmental tree. The samaras hang on some species into the fall, supplying food for the migratory songbird population who specialize in seeds.

DESIGN

The ash has other relatives of note in the garden, notably the forsythia, lilacs, and privet. All of the native species of ash make magnificent specimen trees. There is a choice from the most cold hardy, *F. pennsylvanica* (zones 3–9), to the chubby, fragrant Mexican ash, *F. cuspidata,* which can squeeze itself into zone 8.

The autumn foliage of the ash is very beautiful, going from a dark green to light lemon and then into a misted purple. This color blends particularly well with the crazy oranges and brilliant scarlets of other trees such as the *Acer* species, the maples, in the fall. *F. americana,* the white ash, has a cultivar called *F. a.* 'Autumn Purple' that has outstandingly beautiful, deeper, misted purple foliage.

F. nigra, the black ash, has the same dainty fall colors as the white ash. For the gardener with damp soil, it is an excellent tree with black winter buds.

The green ash, *F. pennsylvanica,* has two interesting cultivars, *F. p.* 'Summit', which has golden fall foliage and a broadly conical habitat, and *F. p.* 'Variegata', a brightly variegated tree, the leaves having a lovely cream white mottling that is excellent in the spring and summer but turns to an indifferent yellow in the fall.

F. velutina var. *glabra,* the Arizona ash, is a comely, small, drought-resistant tree that can grow extremely well on dry, alkaline soils. The leaves and foliage are covered with a dense, gray velvet down. This species grows in zones 6–10, but one could experiment with it in colder gardening zones.

F. quadrangulata, the blue ash, grows to a delightful, high-branching, elegant tree with touches of orange brown in the inner bark. This is highlighted in *F. q.* 'Urbana'. The strange, square twigs add interest to a tree that turns a magnificent yellow-bronze in the fall. This tree, given a rich soil, will mature rapidly to 72 feet (22 m).

The European common ash, *F. excelsior,* has a superior golden clone that has golden shoots and branches and has a clear yellow leaf change in the fall. It is also very remarkable with its golden hue in the winter garden. The *F. e.* 'Pendula' is a strong tree with mounding, divergent, weeping branches tipped with beautiful, black winter buds.

In warmer gardens, zones 7–10, fragrance is delightful from

Fraxinus Species and Cultivars of Merit

SCIENTIFIC NAME	COMMON NAME	ZONES
Fraxinus americana	American white ash	3–10
F. a. 'Autumn Purple'	Autumn purple white ash	3–10
F. bungeana	Bunge ash	4–8
F. caroliniana	Carolinian ash	4–10
F. cuspidata	Mexican ash	8–10
F. excelsior	European ash	4–10
F. e. 'Jaspidea'	Golden European ash	4–10
F. e. 'Pendula'	Weeping European ash	4–10
F. mariesii	Marie's ash	5–10
F. nigra	Black ash	3–10
F. ornus	Manna ash	5–10
F. pennsylvanica	Pennsylvania ash	3–10
F. p. 'Summit'	Golden leafed red ash	4–10
F. p. 'Variegata'	Variegated red ash	3–10
F. quadrangulata	Blue ash	4–10
F. q. 'Urbana'	Urban blue ash	4–10
F. uhdei	Shamel ash	5–10
F. velutina var. *glabra*	Arizona ash	6–10

Lichen on ash, *Fraxinus*. Complex lichen communities such as the greenshield lichen, *Flavoparmelia caperata*, or the gray hammer shield lichen, *Parmelia sulcata*, on the mature bark of any *Fraxinus* species still indicate atmospheric cleanliness.

the June-flowering panicles of the dainty Marie's ash, *F. mariesii*, and the manna ash, *F. ornus*. Ideal for the colder garden is the Bunge ash, *F. bungeana*, well suited to zones 4–8 and flowering in May. This small tree from northern China is remarkable for its down-covered twigs and petioles, adding quiet interest to a winter garden.

Honey locust, *Gleditsia triacanthos,* planted as a cattle barrier in 1823 by an Irishman, John McCrea, who had come from County Cavan, Ireland. They stand as a silent monument to his industry as a farmer.

Gleditsia

HONEY LOCUST

Leguminosae — Zones 3–10

THE GLOBAL GARDEN

In the middle of the "dirty thirties," the common honey locust, *Gleditsia triacanthos,* roared into prominence as a soil saver. The stock market had bottomed out with the dust bowl weather, and even the soil was being blown away and lost. The U.S. president of the time, Franklin D. Roosevelt, put a shelter belt program into place. He ordered the planting of 19,000 miles of trees to stabilize 33,000 farm holdings that had literally "gone with the wind."

The humble honey locust was chosen because of stellar performance in arid India and Africa. This honey locust was a native American child that had, over the millennia, plunged its roots into a diminishing water table. This green baby had checked erosion while at the same time having produced a bumper crop of nourishing, high-quality pods that could be used for food.

The aftershock of the 1930s initiated a unique experiment in Auburn, Alabama. It took place at the Alabama Polytechnic Institute at Auburn (now Auburn University). The common honey locust, *Gleditsia triacanthos,* was picked as the subject of a series of field trial experiments. The scientists tried by selective breeding techniques to improve the tree's edible pod crop. These were then later intended for wider agricultural use as a "new" forage crop. Two quite extraordinary cultivars rose to the top of the honey locust class. These were the Milwood locust, *G. t.* 'Milwood', and the Calhoun locust, *G. t.* 'Calhoun'. The first had pods with a sugar content of 35 percent and the second, 37 percent.

The government that had funded these experiments did not look into their crystal ball. They did not foresee the entire globe changing climatically and threw out the baby with the bathwater. They cut down the trees. The whole experimental protocol was scrapped. Like Nero, as the trees burned, the governments watched the Apollo space project take off.

There are about twelve different species of honey locust, *Gleditsia,* worldwide. These are found in Argentina, South Africa, New Zealand, India, Japan, China, the United States, and Canada. The common honey locust, *Gleditsia triacanthos,* is the most important species in North America. The Chinese honey locust, *Gleditsia sinensis,* is a medicinal tree of traditional Chinese medicine, and *G. caspica* is one of the important agricultural feeding trees in a two-tier agricultural system in the dry areas of the African and Asian continents. The North American common honey locust, *G. triacanthos,* was a member of the ancient virgin forests. It grew side by side for millennia with the beautiful laurel leafed oak, *Quercus imbricaria.* These two trees are now considerably reduced in numbers because of the practices of early settlers.

These people had learned from the local aboriginal population that the honey locust log will not be consumed by termites. So they used them as sill logs for the foundations of their homes and laurel leafed oak as shingles for the roofs of their new homes.

The most startling item about the honey locusts worldwide is the tree's means of protecting its canopy. Around the trunk, about midway, is a girdle of ferocious thorns. In the Chinese honey locust, *G. sinensis,* these thorns are buttressed and branched. This armor gives the tree a mild prehistoric air. The crown of the tree is composed of wonderfully delicate, bipinnate foliage, the length of which can vary with the species. The leaves are burnished in the fall by long mahogany-colored, shining swags of pods. As with many leguminous trees and shrubs, the foliage is toxic, while the fruit is a pleasure to eat.

There are a few tree species that are widely planted and are similar to the honey locust. One of these is the Kentucky coffee tree, *Gymnocladus dioicus.* While the entire pod and the contents of the honey locust, *Gleditsia,* are edible, the pod of the coffee tree, which looks similar, is not. The raw seeds of the coffee tree are very toxic. The aboriginal peoples had, in the past, collected and roasted coffee tree seeds when ripe, thereby detoxifying the seeds, and drunk the infusion as a stimulating drink. Another tree that is related to the honey locust is the eastern redbud, *Cercis canadensis.* The red flowers, and indeed the buds also, are edible and can be used in salads. This beautiful tree is sometimes called the Judas tree. This is because of the Christian oral history surrounding this tree. Jesus, before his death by crucifixion, was betrayed by his best friend, Judas Iscariot, who, anguished by the remorse of his action, hanged himself on a redbud tree. The redbud in question was in Palestine, just a little different from the American redbud. It is *Cercis siliquastrum,* also called the Judas tree. This history also holds a legend that the tree blushed with shame and was forever pink afterward. Judas may possibly have ended his life on a rare *alba* sport of this tree, because *C. s.* 'Alba' and its rosy parent, *C. siliquastrum,* are both to be found in Palestine, or Israel as it has now become.

ORGANIC CARE

The honey locust, *Gleditsia,* can be reproduced by seed and by root cuttings. All of the 12 *Gleditsia* species produce a crop of long seed pods into the fall. In October or November the seed pods can be collected either from the tree or the ground. The outline of the seeds can clearly be seen on the pod. The seeds can be pressed sideways to extract them from the pod. The seeds themselves are quite large, about the size of a bean.

The seeds of the honey locust, *Gleditsia,* have to go through two manipulations before they can be persuaded to grow. The first is scarification. The second is imbibtion.

In the normal run of things scarification happens naturally. A large animal eats the sweet seed pod. The seeds get a good dousing of warm, hydrochloric acid as part of the stomach contents. The seeds then make their two-hour passage down the large intestine and are eliminated in a nitrogen-rich media perfect for growth. The gardener has to mimic this. First the seeds can sit overnight in warm water. This might soften up the seed or testa coat that is so tough in the Leguminosae family. If this does not work, each seed has to be nicked. The nick is an injury that will open the seed coat. The seed is held between the fingers and the more rounded edge is filed with a nail file or a triangular file. The seeds are ready for the second stage.

Imbibtion is a process common to many seeds where the living embryo must be rehydrated or soaked in water. Over a 24 hour period the small seed quadruples in bulk—so much that the seed coat is often separated entirely. The seed is now ready to grow and must be planted without delay into pots or into a nursery row. The seeds are planted with the rounded side facing the planter as they are slipped downward into the soil. The depth should be ¾ inch (2 cm) deep, and the seeds should be spaced 2 inches (5 cm) apart.

Honey locust, *Gleditsia,* seeds can be planted outdoors in this manner in March for zones 5–10. They can be planted outdoors after the middle of May for zones 2–4. For these colder zones one layer of Reemay is placed over the nursery row to protect the growing seedlings from spurious freezing spring temperatures. The seedlings are allowed to grow on for one year, after which time they are easily handled and are readily transplanted the following spring.

The honey locust, *Gleditsia,* likes a rich, moist, fertile soil of pH 6.5–7.5. Above all things, they enjoy good winter drainage. The honey locust will perform well in poor soils provided they are well drained. An underburden of dolomitic limestone seems to invigorate the growth of these trees.

For the gardener or farmer, the common honey locust, *Gled-*

itsia triacanthos, will grow climatically wherever a crop of corn can be grown. Usually this is zones 3–10 for this continent. Field trials should also be conducted on it in zone 2.

If the honey locust, *Gleditsia,* seeds are not to be planted immediately, they can be stored for several years. They should be wrapped in brown paper and placed in a glass jar and put into a refrigerator at 32–45°F (0–5°C). The imbibtion process for stored seeds will be several hours longer and may take up to 30 hours for rehydration.

Gleditsia triacanthos, the common honey locust, is remarkably disease free. This tree can reach 140 feet (43 m) with maturity. This species is known to live for at least 200 years. The actual age of the mature species in the virgin forest is unknown. The thornless cultivars of the honey locust seem to have exchanged the thorns for infection. *G. t.* 'Moraine', and *G. t.* 'Shademaster' are more susceptible to the pod gall midge, *Dasineura gleditschiae,* which affects the growing tips. However, this disease can be prevented by proper spacing. The cycle of this disease can be broken by planting the trees 21 feet (7 m) away from any building that would harbor overwintering eggs. Occasionally, the yellow leafed cultivars, *G. t.* 'Sunburst', are attacked by the honey locust plant bug, *Diaphnocoris chlorionis.* This happens in the spring when the young, sweet, yellow buds are breaking. A mild foliar spray of neem tree oil will control this.

MEDICINE

Gleditsia triacanthos is a medical tree of North America. *G. sinensis* is the corresponding medical native species of China. Both are members of the pea or Leguminosae family.

The medicines of the tree are to be found in the bark, the thorns on the bark, the seed pods, the sweet pulp inside the pods, and the ripened seeds in a dried state.

The pods of both trees are of agricultural interest. There are additional biochemicals in the trees that could be of use to the

The mahogany-colored, ripe, leguminous pods of the honey locust, *Gleditsia triacanthos*

The barrier of spines on *Gleditsia triacanthos,* a cunning means to protect the delicious foliage of the crown of the tree

pesticide industry, the dairy industry, timber protection in the forestry industry, the health food industry, cardiac research, and in cancer research and prevention.

The Cherokee aboriginal peoples used the fresh, sweet pulp from the near-ripe pod of *G. triacanthos* for the treatment of dyspepsia. In addition, a whole pod extract was used in the treatment of measles.

The Delaware aboriginal peoples used a bark decoction of the common honey locust, *G. triacanthos,* as a tea. This was used for asthma and the coughs of whooping cough. They also used the tea in the treatment of various blood disorders such as stroke.

The Chinese traditional medicine used the Chinese honey locust, *G. sinensis,* in a similar manner to the Delaware aboriginal peoples, for the treatment of sore throats, asthmatic coughs, and strokes.

The Fox aboriginal peoples also used a bark decoction of the honey locust, *G. triacanthos,* as a tea to treat colds and flus. This tea was also used in the management of smallpox.

The native peoples of South Africa, especially around Lesotho, used the Caspian honey locust, *G. caspica.* The fruit pulp from ripe seed pods was used to rid the lungs of phlegm in the treatment of catarrh. They also collected the seeds of the same tree and sun-dried them. These they ground into a very fine powder. This powder was then taken as a form of snuff.

Interestingly, the Chinese traditional medicine used the multi-spines of Chinese honey locust, *G. sinensis,* and the North American aboriginal peoples used the single spines of the common honey locust, *G. triacanthos,* for the same purpose. The spines of these two trees are counterirritant. They are highly antibacterial and antifungal. Unlike all other thorns, if used as a tissue probe, they will go in and come out cleanly without causing infection.

The Chinese used the multi spines of the Chinese honey locust, *G. sinensis,* in the treatment of carbuncles and other infected skin lesions to disperse the toxic matter of infection. The North American aboriginal peoples used the spines or thorns of the common honey locust, *G. triacanthos,* as a sterile surgical instrument to conduct a form of subepidermal surgery of the skin. Five thorns of the common honey locust, *G. triacanthos,* of identical diameter were attached in a straight line of five tips to a wooden handle to make an instrument. This was guided by a knife to apply medicines by surface puncture into the subsurface of the skin. This surgical procedure was used in the treatments of neuralgia and rheumatism.

The medicine that was applied on the tips of the spines was a moistened mixture of the dried gall of a black bear and charcoal as a carrier. The charcoal was obtained from eastern white cedar, *Thuja occidentalis,* or the native hazel, *Corylus americana.* In the case of the latter, our native hazel, a completely new class of anticarcinogenic chemicals have just been discovered. These are called taxanes. They appear to be manufactured by an as yet unidentified fungus living in the hazel wood. The Chippewa aboriginal peoples used this surgical technique and native medicines in their daily rounds. The patients seem to have been cured but always continued to carry the charcoal's black pigmentation underneath the skin's surface. An excess of this form of pigmentation in the aboriginals was noted by the pioneers in the first white settlements.

Two chemicals, fustin and fisetin, are to be found in the heartwood of the common honey locust, *G. triacanthos.* Fisetin, in particular, and its daughter compound, fustin, have potent anticarcinogenic action. Fisetin has the ability to block or to inhibit a whole family of killer chemicals called aflatoxins. Fustin and fisetin are probably common to all *Gleditsia.*

A local anesthetic called stenocarpine is extracted from the common honey locust, *G. triacanthos.* It is also thought that cocaine can be extracted. Usually cocaine is extracted from *Erythroxylon coca* of the tropics. Nitrogen fluctuations in the tree's growing conditions may induce enzymic changes, much as happens in the basswood, *Tilia,* species. Cocainelike compounds may be produced by the trees as an insect killer to enrich the soil with nitrogen, much as is practiced by the insectivorous plants, the pitcher plant, *Sarracenia purpurea,* who prefer their insects whole.

ECOFUNCTION

In the world of trees the honey locust, *Gleditsia,* is no ordinary tree. Its strategies for species survival are unusual, to say the least. The tree employs two antifeeding alkaloids, triacanthine and gleditschine, to protect the leaves. The full canopy is protected by a crown of thorns. The seed crop is protected in a very clever way. The tree depends on large creatures to disperse the seed crop to ensure survival. Mixed with the seeds is nature's temptation, sugar, in the form of a wonderfully sweet pulp. Most animals have a sweet tooth and will gladly eat the seed pods and disperse the seeds in the usual way. But sweetness in the seed crop brings its own share of worries for the tree, fungi. The entire kingdom of the fungi, too, like the easy road and will gladly accept sweets. But fungi spell disaster for *Gleditsia,* because fungi will not only rot the seed crop, it will also make it so toxic that no sane animal will come close to it. The honey locust circumnavigates this little problem with ease. Instead of being a grateful legume of the Leguminosae or pea family that dines on nitrogen with the help of rhizobium bacteria adhering to its roots, as is normal for this family, the honey locust manufactures antirhizobial compounds to get rid of these bacteria. This leaves the soil around the tree somewhat sparse in nitrogen, which also means that fungi do not find it a tempting place to live. To cap this, the tree produces a nuclear missile in the form of fisetin, just in case there is a wandering minstrel form of fungus out there. Fisetin will entirely destroy the ability of the fungus to establish itself in the mature seed pods by blocking the production of a cell metabolite called aflatoxin that would make the seed toxic to animals. Fisetin does this by destroying the DNA synthesis in the fungal cell. This is the ultimate weapon for all living creatures, the ability to destroy its encryption for life. And, that is why all the honey locust, *Gleditsia,* species will be useful in cancer research.

The pinnate leaves of the common honey locust, *G. triacanthos,* have the unique ability to move with purpose. Independent movement is rarely seen in the plant kingdom. It is found among

Opuntia humifusa, prickly pear cactus, which shares with *Gleditsia triacanthos,* the honey locust, a love of dry places

carnivorous plants and among trees like the mimosa tree, *Albizia julibrissin.* The pinnate leaves fold into a closed position like a bivalve when the light conditions are reduced or in darkness. This phenomenon might be tied to a highly specialized nitrogen cycle in the tree itself.

The girdle of thorns found on the *Gleditsia* species are produced from the inner cambium of the tree as it grows into maturity. The base of the thorns is wide, making them tough, sharp, efficient barriers to porcupines, opossums, and other mammals that would otherwise forage on the tree. This barbed wire system protects the canopy as it matures. Once the tree has attained maturity for its given site, the 8-inch (20 cm) thorns are shed and the bark assumes a regular outline.

In many areas of the southern United States the common honey locust, *G. triacanthos,* crop has been used in mixed farming. The pods have been used as a forage crop in the fall. In some instances the pods are collected and stored dry, much as hay is stored. This is used as winter feed for heifers. The pods are fed to sheep, who appear to digest the entire pod. Cattle will feed on them and excrete the seeds. In many areas of the Carolinas the bean pods are ground and mixed with other feed, or they are mixed with alfalfa meal as part of the dairy ration. Such feed is said, by dairy farmers, to increase milk production in their cows. This mixture will also reduce the chance of having the highly toxic 4-hydroxylated aflatoxin B derivative, which is occasionally found in the milk of cows fed molding meal by default. Pathogenic fungi grow well in hot, humid summers.

The honey locust, *Gleditsia,* is excellent bee pasture. In the late spring the fragrant flowers are a good source of nectar for honeybees. In zones 3–5, the locust is one of the last trees to leaf out and immediately produces racemes of late flowers in perfect timing for the end of the first year's brood rearing.

The common honey locust, *G. triacanthos,* is also the host tree to the dainty silver-spotted skipper butterfly, *Epargyreus clarus,* and its clan of golden-banded and golden-spotted skippers. These butterflies have adapted to modern suburban living and quite often give spectacular aerial displays of flight. The dreamy dustwing, *Erynnis icelus,* is also thought to use the honey locust tree.

The common honey locust, *G. triacanthos,* has learned to share a habitat with termites. These insects love to dine on wood. The wood of the locust is infused with fustin and fisetin biochemicals that makes the wood termite-proof. The other honey locust, *G. aquatica,* of North America might also have similar potent, natural insecticides in the roots, wood, and leaves.

BIOPLAN

If corn can be used for ethanol production, there is no reason why the high sugar producing cultivars of *G. triacanthos,* the common honey locust could not be used for fuel. The honey locust could be used to produce "green power" for many North American farms. France has a project, too, on the drawing board, based on fuel production from the sugar energy of locally grown sugar beets.

The *Gleditsia* cultivars *G. t.* 'Milwood', *G. t.* 'Calhoun', and *G. t.* 'Majestic', all high sugar producers, could be selected and harvested for biomass to produce ethanol fuel. Farms or small towns of North America could have "set-asides" for "green energy" conversion for their own use. The technology, already in place in North America for conversion of corn sugars into alcohol, would be easily adjusted for processing *Gleditsia* sugars.

The high sugar producing cultivars of the honey locust can be used as part of an agricultural bioplan. The trees could be used in two-tier production as hedging or field islands where cropping is carried out on a normal basis. The pod crops are an additional feed bonus. This pod crop can be free grazed or winter stored or processed as silage or mixed with alfalfa as a ration.

The common honey locust seeds were eaten in the Deep South. They were soaked overnight and then cooked much as marrow peas are cooked in England. These seeds are an excellent source of vegetable protein called avenin or vegetable casein. This protein is similar, and indeed related, to soya beans. In addition, a locust bread was well known in the southern kitchens. The entire pod was ground into a sweet flour used in a ratio of one-third locust flour to two-thirds corn meal. This was used as a baking flour. The best cakes are said to be made from the flour of the *G. t.* 'Milwood' cultivar. A drink was made from an extract of the seed pod. The locust pods were also added to persimmons to make the famous persimmon beer of the South.

City life and heavy industry appears to be well suited to *G. triacanthos* and all of its cultivars. These species will tolerate a higher level of pollution than do most other trees. They absorb increasing amounts of carbon dioxide as carbon banks in otherwise unhealthy environments. They are atmosphere scrubbing trees that can grow well on the margins of safe habitat, particularly in city environments. Such trees will reduce the particulate matter and reduce childhood asthma associated with suburban environments. There are also some improved, disease-resistant cultivars coming into the market. These are *G. t.* 'Imperial' and *G. t.* 'Moraine'. Both of those species will be important carbon-banking trees helping to clean industrial areas. A number of other patented cultivars are also on trial, *G. t.* 'Halka' being one example.

Termite populations will track with the heat of global warming. *G. triacanthos* produces termite-proof wood. Lumber and forestry companies should initiate "set-aside" plantings of the honey locust as a source of this lumber. Presently lumber is being pressure-treated with extremely toxic arsenic and chromium metals as poisonous, chemical shields to termites. But arsenic, even such low concentrations as ten parts per billion (ppb) in water, seems to function as an endocrine disrupter, according to recent research.

In the northern gardens of North America, the sunburst honey locust, *G. t.* 'Sunburst', is being increasingly planted. This beautiful little cultivar is extremely drought-tolerant. In times of water shortage, which seem to be on the increase, especially in cities, this tree is unaffected. It will cast dappled shade that prevents scorch of both grass and plantings nearby. The ability of this tree to reduce the rate of evaporation will become more important as freshwater comes into short supply. *G. triacanthos* and all of its cultivars are xerophytic trees and can be part of a drought-tolerant landscape.

DESIGN

The native *G. triacanthos,* the common honey locust, is an elegant tree, growing tall and slim in a northern climate. It breaks into leaf in June and almost immediately into long racemes of fragrant white flowers not unlike wisteria. The airy crown is made up of either pinnate or doubly compound leaves. As the crown matures, it becomes flat topped with a rich, pea green foliage from spring, changing into the gold of fall. The swags of long shining mahogany-colored pods become a focal point of the late fall garden.

Of the various garden cultivars the sunburst, *G. t.* 'Sunburst' (syn. *G. t.* 'Inermis Aurea') is the most common. Its startling lemon sunshine foliage is a relief from the starkness of winter

The lupine and the *Gleditsia triacanthos,* honey locust, share the same flower form and are both members of the Leguminosae or pea family.

and seems to herald spring in a northern garden. The sunshine can be captured and amplified by any of the yellow, double-flowered narcissus or the lemon and orange late tulips like the Darwin tulip, *Tulipa* 'Hans Mayer', or the late-flowering tiny tot, *T. batalini* 'Apricot Beauty'.

For the smaller garden, there is an old weeping cultivar, the Bujotii locust, *G. t.* 'Bujotii' (syn. *G. t.* 'Pendula'). This is a small tree with more delicate pendulous branches. The small, ever-blooming daylily, *Hemerocallis* 'Stella d'Oro', makes a perfect, carefree companion to this honey locust cultivar.

There is, in addition, a honey locust cultivar, *G. t.* 'Ruby Lace', with rich ruby spring foliage. The ruby color is stronger in the colder gardens of zones 3–5. From June into July the red-purple mist of foliage can be used as a foil to carry the color throughout the garden. The burned reds and maroon colors of the hybrid lupines called Russel's hybrids, *Lupinus polyphyllus* 'Russel Hybrids', combine and thread these colors throughout the garden in a rich form. Both the tree and lupine share the same family ties and therefore similar growing conditions.

Gleditsia Species and Cultivars of Merit

SCIENTIFIC NAME	COMMON NAME	ZONES
Gleditsia sinensis	Chinese honey locust	3–10
G. triacanthos	Common honey locust	3–10
G. t. 'Bujotii' (syn. *G. t.* 'Pendula')	Weeping honey locust	3–10
G. t. 'Calhoun'	Calhoun honey locust	3–10
G. t. 'Imperial'	Imperial honey locust	3–10
G. t. 'Majestic'	Majestic honey locust	3–10
G. t. 'Milwood'	Milwood honey locust	3–10
G. t. 'Moraine'	Moraine honey locust	3–10
G. t. 'Ruby Lace'	Ruby lace honey locust	3–10
G. t. 'Sunburst' (syn. *G. t.* 'Inermis Aurea')	Sunburst honey locust	3–10

The spines of *Gleditsia triacanthos*, which have an unusual and unexpected medicinal action

A young allée of black walnut, *Juglans nigra,* in the spring with butterfly daffodils

Juglans nigra

BLACK WALNUT

Juglandaceae

Zones 3–9

Black walnut, *Juglans nigra.* The fall colors of the black walnut, especially in an allée, create an arcade of gold.

THE GLOBAL GARDEN

The black walnut, *Juglans nigra,* is the mistress of the North American forest. She is as capricious in growth as she is bountiful, though few realize her full potential. She eludes those who dearly wish to grow her and plunks comfortably into the land of many a farmer who is careless of her charms—that is, until the word "stumpage fee" is spoken: $30,000 is not unusual for a black walnut.

There are 15, perhaps even 16, members of this extraordinary family in the global garden. If the great forest slopes of the Pou-Nam near Lai-Chau bordering on Vietnam, China, Myanmar, and Laos are searched botanically again, another lurking member of this family might emerge. The Vietnamese are presently looking for it. The Juglandaceae or walnut family resides all over the known globe. The most ancient of these trees, from an evolutionary point of view, lives in China. In North America there are six *Juglans* species of this family. Two of them are of particular interest. The black walnut is one of them; the butternut, *Juglans cinerea L.,* is the other.

The black walnut, *Juglans nigra,* is a tree of legend. The very name, *Jovis* and *Glans,* means Jupiter's nut. Now Jupiter was the chief Roman god who was also the husband of Juno. He was god of light and in passing was also god of the sky, weather, of the state, its welfare, and of course, its laws. This looks after

Juglans. The *nigra* means black. Every person who has handled these nuts in dehusking for planting will know the full meaning of this word because the person's hands will have been rendered black for days afterward, the black fading to an ugly brown a week or so later.

The importance of the Juglandaceae was evident in the Roman Empire long before Christ. It seems that walnuts were a favorite food three thousand years before that. They were haute cuisine for the Babylonians.

Even King Solomon, in the millennium before Christ, had orchards of these nuts. As king of Israel his wisdom was boundless. He took short breaks in his cerebral career by sitting in his nut grove and eating their flesh in the fall. He was wise even in this, because nuts are a brain food. And, somehow, the Persians had already known this for a long, long time.

Parallel in time the aboriginal peoples of North America highly valued *Juglans nigra* as a food. The trees were owned separately from the land on which they grew. Even today, this is a unique concept in land management. The flesh of black walnuts was a protein-packed winter food carefully hoarded in tall, stilted buildings. Pound for pound, the flesh of these nuts and the flesh of beef have an equal value. Plus, as the aboriginals knew quite well, the nuts were endowed with a little sedative that promoted a nice snooze in the dull depths of a longhouse winter.

Up to about one hundred years ago the black walnut grew in the cotton belt, the corn belt, and around the Great Lakes, tracking a little northward within the Canadian river valleys. The timber, which is easily carved, was a choice wood for the Arts and Crafts movement at the turn of the last century. That, multiplied by a stunning ignorance of the true value of this tree, has now made the tree scarce and rather a novelty with its disease-resistant timber much sought after as a source for dark chocolate-colored veneer.

Strange though it may seem, black walnuts are a very much undervalued resource in North America. A single tree planted now will pay for the university education of one's grandchild. Few investors realize that the addition of high-quality timber such as black walnut, or *Juglans nigra*, to any long-term investment portfolio will bring in dividends greater than any mutual funds.

ORGANIC CARE

There are three limiting factors to the growth of the black walnut, *Juglans nigra*. These are temperature, soil depth, and apical damage in the young tree.

Extremes of heat in the tropical climate of Florida are said to damage the blossom at the time of fertilization. In a similar fashion, extremes of cold in the north of its range of zones 3–4 will damage both the blossom and the growing apex of the tree. Most often this particular damage occurs just after the tree has broken dormancy, and elongation has begun. If there is a killing frost at this time, it will severely damage the apex, whose function is taken over by the first dormant bud down the stem. This will require subsequent corrective pruning. Apical damage is common in areas where the deer population is high and deer either eat or damage young walnuts when rutting. It is also common in areas that are prone to severe freezing rain. Severe winter temperatures of zones 3–4 can also cause apical damage. If this kind of injury of the apical area occurs, the tree is not useful as a timber tree.

Black walnuts require a deep, rich, fertile soil. A shallow soil is not compatible with healthy black walnuts. The soil should have a neutral to slightly alkaline pH that is anywhere from 6.8 to 7.2. Seedling black walnuts should be planted in their first or second year of growth. Great care must be taken not to damage the young growing tap root at this time. The taproot is both long and brittle. The tip should have a clean, white appearance similar to the flesh of an onion in its crispness. The tree root is sprayed with water and covered with damp burlap while the hole is being dug. The tap root is then measured. The hole for the black walnut should be at least 6–12 inches (15–30 cm) deeper than the length of the tap root. Aged sheep or horse manure at least one year old should be mixed into the planting hole. Two gallons (7.6 l) of either of these manures will suffice. It should be well mixed with the soil until the manure is no longer evident. If the soil is a little on the acidic side, 1 quart (.9 l) of dolomitic lime should also be added to the soil and mixed. In more shallow soils or poorer soils phosphate can be added in the following manner. Leftover soup bones, a minimum of 5 pounds (2.3 kg), are placed in the hole below the tap root and covered with soil. The tree is carefully planted, fitting comfortably into the hole. It is well watered and a shallow rainwater catch is left around the tree that is 2–3 feet (.6–.9 m) in diameter. The bones supply a slow-release phosphate source that lasts for 5 to 10 years.

Walnuts are planted in January to February in zones 7–9. They are planted in March in zones 5–6 and in April in zones 3–4, or as soon as the ground can be reasonably worked without the hole filling with water from the hydrostatic pressures of the spring water runoff.

In cold gardens in zones 3–4, the planting site should have some consideration. The young walnut tree should be planted where it has deep winter shade. This can be on a north-facing slope or in a northeasterly position near a hedge or house. A cold site will delay growth in the spring, which will in turn save the young tree from late frost damage. In addition, frost-hardy cultivars are available that have a shorter growing season. Because global warming is bringing erratic climatic behavior in its trail, gardeners in zones 5 and 6 could look to planting more frost-hardy cultivars.

Tree trials of black walnuts in Carrigliath show interesting results. Nuts from the same crop were planted in poor shallow soil. These nuts were mulched with dolomitic stones. Other nuts were planted in enriched shallow soil. In 20 years the nut trees on poor soil had grown to 3 feet (.9 m) tall, showing no sign of bearing. The nuts in the enriched soil were 18 feet (5.5 m) tall and were beginning to bear. The trees on the poor soil never grew beyond the height restrictions for feeding deer and had to be constantly protected during this time.

In northern climates the black walnut has a considerably

The red female and green male flowers of a *Juglans nigra,* black walnut. They are self-fertile. Such species are said to be *monoecious.*

The purple-flowering raspberry, *Rubus odoratus,* often found growing near mature *Juglans* species because birds and berries go together with a wild feast of nuts

shorter growing season, 140 days and less as you go farther north. This is in contrast to the 280 days of lush growing of zones 8 and 9. Consequently, in the north, black walnut trees need a longer time to produce wood that will support nut production. This may be from 12 to 20 years. Trees in the south will begin bearing at 7 years. However, despite these differences black walnuts are equally faithful to their task of nut production in a warm or cold climate. This is important for the black walnut industry.

Black walnut trees are propagated by grafting and nut stratification. The grafting techniques take some skill and are methods used by professional plant propagators who want to propagate a particularly worthy cultivar. A method called cleft grafting is used in which the scion nutwood cultivar of choice is cleft grafted into a mother seedling of black walnut, node to node and tied with beeswax or banding to keep the cambium tissue matched and to keep it from desiccation. In zones 3–5 an additional Reemay or burlap wrap will protect the graft union from cold damage throughout the first winter. Bud grafting is used to a lesser extent.

Black walnut trees will lose all their leaves in one day after the first killing frost. The nuts, either singly or in clusters, will remain on the tree. Within a week or so, these too will fall off the tree, all in one day beginning at dawn and ending almost at dusk. Black walnuts should be collected at this time. Within two to three days of this event the nuts produce a chemical calling card to the surrounding squirrel population and "the rest is husk." The cunning of squirrels and their love of the black walnut should never be underestimated by those who wish to store these nuts for spring germination.

The husks can be removed from the nuts while they are still green after collection. The seeds can be put into plastic bags and kept at 37°F (3°C) for 90 days (zones 7–9) and up to 120 days (zones 3–6). The nuts are then planted 2 inches (5.0 cm) deep and 8 inches (20 cm) apart in the rows of a nursery bed. A 3-inch (8 cm) mulch of sawdust should be placed over the nursery beds of zones 3–6 to prevent the heaving of the nuts to the surface at times of freezing and thawing. All beds should be well labeled, as black walnuts may take from one to two years to germinate.

For larger quantities of nuts, they can be buried 2 feet (60 cm) deep in soil or mulch outdoors during the winter season and planted in the spring as early as the ground can be worked. Excellent protection from curious and hungry squirrels must be provided.

Young walnut whips have to be protected from white-tailed deer in the fall. As winter begins to threaten, the bucks' antlers become itchy. Bucks will search out members of the *Juglans* family to scratch these antlers until they are finally removed altogether. This causes tremendous damage to the young trees. This can be avoided by loosely tying a bamboo stake to the young tree with twine. The top of the stake should be above the apical bud to prevent that from being nipped off, and the bamboo stake will prevent the buck from whipping the stem backward and forward in his frantic attempts to itch.

Black walnuts are pruned in the fall. Whatever the zone, the tree should be fully dormant before pruning is attempted. Nuts should be pruned using a sterile technique. The cutting tool is soaked with a 2 percent solution of household bleach in water, 1 part household chlorine bleach to 50 parts water.

MEDICINE

There are a number of important medical compounds found in the black walnut. These are juglone and ellagic acid together with tannic acid and a fixed oil among other unclassified compounds. Juglone and ellagic acid form the basis of the black walnut's being an important medical tree.

Ellagic acid is a potent antagonist of the mutagenicity of several harmful aromatic hydrocarbons. This acid is a prototype of a new class of cancer-preventing drugs. It is also used commercially as a hemostatic agent that prevents excessive bleeding.

Juglone has sedative properties and is a potent inhibitor of fungal growth. It is also used in the chemical industry as a pH indicator. This is a compound that by its color will tell if a solution is an acid or an alkali. It is also the source of natural brown F, a natural dye that is also known as nucin in Europe.

In the warmer regions of the world human fungal infections are endemic. This was the case for early Roman and Greek societies. The bark and the fruit of walnuts were used to treat cases of fungal infections of the skin.

In Pakistan there is a tradition of using the stem bark of walnuts as chewing sticks, an ancient practice. The stick is chewed until the tip becomes a brush that is used as a toothbrush for disturbing plaque and preventing its growth. While this process goes on, the antifungal compounds in the wood are leached out into the saliva in microquantities and help to readjust a healthy balance of fungi and bacteria in the mouth.

Black walnut oil extracted from the nuts is very high in one of the essential fatty acids, linoleic acid. This fat is necessary to healthy brain development in young children and the maintenance of cognitive power as the adult ages. Few foods are so high in this acid. These nuts as a flour should become an essential food in the vegetarian diet. They are a major part of Russian provincial cuisine.

Lost and Found

The Carpathian walnut, *Juglans regia,* cropped over 200 pounds of nuts in the fall. These were bagged and buried 2 feet (60 cm) deep within a giant wire basket to keep squirrels at bay.

A month later I was stopped dead in my tracks. On the pathway lay the husk of a nut—a *J. regia* nut! The red squirrel sat overhead, filing and dining with salivary delight.

The squirrel had used the bars of the basket as a lever system. An exquisitely delicate little entry hole could be seen. But, this is not a sad story. I managed to save half of the nuts for my millennium project . . . IQ for IQ, the Ph.D. went to the squirrel!

It should not go unnoticed that when deer rake the sides of sapling black walnuts, they are releasing just sufficient ellagic acid to clot the bleeding area of the antler stem on the crown of their heads. In this manner they are directly using the medicine of black walnuts.

Black walnuts naturally grow by the banks of rivers and on floodplains. The husks of the nuts have their highest content of juglone in the fall just prior to the time the external world goes into dormancy through natural sedation. Juglone release helps the sedation of both animals and fish in these riparian areas and thus functions as a natural medicine. The greater the nut load of a tree, the greater the dormancy effect.

The Seneca nation, in more modern times, used black walnut in the native medicines for the treatment of blood. It is to be assumed that this is a treatment for anemia. A handful of the growing tips of wild asparagus, *Asparagus officinalis,* a natural diuretic, was boiled with a handful of bark of black walnut, *Juglans nigra,* in 12 quarts of water (11.4 l). It was reduced down to 4 quarts (3.8 l). Brandy was added along with 5 lbs (2.3 kg) of brown sugar. One tablespoon was taken before each meal.

For the treatment of headache, bark of black walnut from young growth from the east side of a tree was scraped upward. It was then wet with warm water and a pinch of salt. This was used as a direct poultice on the region of the head with pain.

Black walnut has also been used by the Seneca for rainmaking. Bark that had been hit by lightning was placed in a cup of water for a few minutes. Rain ensued in two days. The bark was saved and dried for future use. This rainmaking practice should not be discounted out of hand because the electron power and voltage in an electric strike is beyond the capacity of man. The resulting fusion could well produce some very interesting chemical products of polymerization.

The bark, hulls, twigs, and leaves of black walnut produce a first-class brown dye. It is fast, and no mordants are required. Concentration affects color depth. Wool, wool-silk, silk, and plant fibers can be dyed. This dye was also used in Rome to tint the silver of aging.

ECOFUNCTION

Within the forest community the black walnut or *Juglans nigra* is among the most important feeding trees for quality protein on the eastern portion of the American continent.

The nuts produced are food for squirrels, mice, and all of the animal chain that feeds on them. Wild boars can exist on the mast from these nuts. Indeed in the 1920s and 1930s when food was scarce, the hog farming in the southern portion of the continent depended on the mast from these nuts, as did the laying hens after the shells had been cracked by pigs. During the winter months when meat was scarce, black walnuts fed the African-American population of the Deep South, and of course they were always a staple for the aboriginal peoples.

Black walnuts are associated with an increased butterfly population. The young yellow leaves produced by an elongating meristem are both food and home for some early hairstreak butterflies.

Black walnuts are solitary. They produce their own chemical weed killers by an ingenious device to ensure that they remain so. A number of daughter chemicals are produced throughout the plant in the roots, stems, nuts, and leaves. These are fungicides; they inhibit the growth of root mycorrhiza of any seedling nut that dares to grow within the leaf canopy of the tree. This keeps the tree a solitary performer. This situation is aided, of course, by the squirrel population, who busily roll the nuts with their paws into their earthy winter's larder. This, incidentally, is the correct position for the planting of the nut . . . on its belly.

With time another ecofunction of the black walnut has evolved that will put this tree into the forefront of the plant kingdom as an antipollution device. When fossil fuels are burned, they produce a number of polycyclic aromatic hydrocarbons. Some of them, like benzene, are directly tied in with increased rates of various cancers. These hydrocarbons can polymerize in the air and form macromolecules called arenes, which were found by Japanese scientists to be responsible for pollution-triggered lung cancer. In the leaves of the black walnut there is an agent, ellagic acid, that directly inhibits the mutagenicity of the ultimate carcinogenic metabolite, benzo-α-pyrene, from fossil fuels. It is estimated that the average urban infant will ingest 110 nanograms of this toxin. This is equivalent to the child smoking three cigarettes on a daily basis. Thus, black walnuts, planted in urban areas, act as pollution sweeps to clean the air, in suburbs, cities, and towns.

BIOPLAN

It might come as a surprise to many North Americans that the bioplanning of nut trees is a truly ancient practice on this continent. Early accounts record that, in 1607, Captain Gilbert described the trees at a point on the Maine coast as "ocke and wallnutt growing in a great space assonder on from the other as in our parks in Ingland and no thickett growing under them."

These pioneers were among the first to document the effects of the horticultural practices of the aboriginal peoples, who had produced great, open, parkland-style savannahs ideal for hunting. These were either copied or improved by the famous English landscape architect Capability Brown, a century and a half later, for the estates of the rich gentry.

North America had already seen a great city built around the top-class protein produced by the black walnut. This city had a greater population than London, England. It was Monk's Mound at Cahokia on the banks of the Mississippi. The citizens of this city had already built an earth mound to rival the great pyramids of Egypt well before the Crusaders were preparing to shed their blood for Christian wars against the Muslims.

The aboriginal peoples maintained an open canopy, which is essential for the black walnut's vigor, by means of fire. They flash-fired the dead, long grass in November and April. These fires produced ash high in potash and calcium, essential macroelements for nut production. The ash also acted as a high-pH pesticide around the walnut trees, which kept them healthy. In addition, the black walnuts were groomed for noxious insects by two species of birds that were once common on this continent and are now extinct. These are the heath hen, *Tympanuchus cupido cupido,* and the passenger pigeon, *Ectopistes migratorius.* The flash fires maintained the open setting for various species of nut trees and grassland.

Every farm and smallholding should have a nuttery attached. Black walnuts should be planted on a grid 40 feet by 40 feet (12 m by 12 m). The nut trees can be interplanted with woody nurse crops that fix nitrogen by rhizobia on their roots. Yellowwood, *Cladastris kentukea,* zones 4–9, or Siberian pea shrub, *Caragana arborescens,* zones 3–9, can be used as a nurse crop. Any nitrogen-fixing clover except alfalfa can be sewn in between the trees. Nitrogen swapping will take place between all of these species for the greater benefit and the health of the nut trees.

A resistant butternut, *Juglans cinerea* 'Batesii', of particular interest to nut culture in North America. It is closely related to the black walnut, *J. nigra.*

Farmland that has black walnuts gives a significant increase in returns. It can be considered to be two-tier farming. This has long been practiced in France, in the Dordogne region, where an acre of land and a nut tree have equal agricultural value. In this area of France the nuts are harvested from the hedgerow plantings. Some of the annual crop are used in "La Vieille Nois," a much sought after nut liqueur.

As an agricultural crop, black walnuts are unique among other nuts, because they hold their flavor even if cooked or processed. Thus the food industry has a huge demand for a constant supply of these nuts that is not being met.

Because benzo-α-pyrene is present around schools due to heavy traffic and because it is directly tied in with leukemia in children, planting black walnut trees in school yards should be part of a bioplan for the health protection of all schoolchildren.

Gas stations, truck stops, and roadways, too, should enjoy the shade of black walnut, as the leaves busily neutralize the offending carcinogens from fossil fuel emissions. These trees can be part of the cure, buying time for research and development to find other answers to pollution problems.

In the suburban garden black walnuts should be planted the greatest distance away from the vegetable garden to a maximum of 40 feet (12 m). The roots, as they grow in size, produce greater amounts of juglone, which will affect the growth of all the Solanaceae family (potatoes, tomatoes, peppers, ground cherry, tobacco), the Ericaceae, or the heath family, blackberries, and members of the red pine family. Black walnuts should be planted as a specimen tree at a point nearest to the roadway to dilute traffic vapors or near the family garage for the same purpose.

Children should be encouraged to play with these nuts as they have done for millennia, rolling them in their hands like worry pieces and thus spreading the juglone and ellagic acid on little hands that need all the protection nature can give when television separates them from such physical contact with the natural world.

DESIGN

In a spring gale of 1822 the largest known black walnut tree was blown down. It had a trunk girth of 36 feet (11 m). It soared 80 feet (24 m) to the first limbs! In all of North America there is just one place left to find trees of this caliber. It is the 60-acre (24 ha) forest called the Indiana Pioneer Mother's Memorial Forest south of Paoli, Indiana. It is the only virgin forest of *Juglans nigra,* black walnut, on the North American continent that remains.

The black walnut, or *Juglans nigra,* is a wind-pollinated tree, and as such it raises its lofty canopy into the air in a determined fashion. The tree has the character of haute couture with a long display of male catkins that develop from the previous year's outer nodes. The orange-tinged female flowers appear with the advent of spring foliage and swell during the summer to the largest nut to be found in the north temperate forest. In the fall when the tree's duty is over the trees are simply laden with nuts. These change the design appearance of the tree into a more drooping companion. The fall colors of black walnut, especially in an allée, create an arcade of gold. The filigreed long leaf with its companion 15–22 leaflets drops to the ground, and silence reigns again around the mature trunk with its rough furrows of deep intersecting ridges.

Black walnuts should be used either as single specimens, in an allée, or in nutteries. As a single specimen the *J. nigra* 'Laciniata' has more finely developed foliage and gives a delicate feminine appearance. The following cultivars should be grown for nuts: *J. nigra* 'Elmer Myer', *J. n.* 'Emma Kay', *J. n.* 'Sparks-147', *J. n.* 'Hay', and the older variety of *J. n.* 'Thomas'. These will change as superior cultivars are continually being produced by enthusiasts and nut experts alike.

Juglans and Cultivars of Merit

SCIENTIFIC NAME	COMMON NAME	ZONES
Juglans nigra	Black Walnut	4–9
J. n. 'Emma Kay'	Emma Kay black walnut	4–9
J. n. 'Hay'	Hay black walnut	4–9
J. n. 'Laciniata'	Laciniata black walnut	4–9
J. n. 'Sparks-147'	Sparks-147 black walnut	4–9
J. n. 'Thomas'	Thomas black walnut	4–9
J. n. 'Carrigliath'	Carrigliath black walnut	3–9

The black walnut, *Juglans nigra,* one reflection of a green planet

Magnolia acuminata, cucumber tree, the tree saved from the slice by an hour's wait

Magnolia acuminata

CUCUMBER TREE

Magnoliaceae

Zones 4–10

THE GLOBAL GARDEN

Magnolia acuminata, the cucumber tree, upstages the rose for genetic heritage in Canada. Fossil fragments from the cucumber tree dated to be around sixty million years old were found in Alberta, whereas the rose, *Rosa carolina,* is a mere thirty-five million years of age. To have hosted the *Magnolia acuminata,* the Red Deer River region of Alberta had to have once been a lush, humid, tropical oasis. But times change, especially for this Pygmalion of trees. The fair lady is vanishing as we sit and eat our daily bread. Along with the tulip tree, *Liriodendron tulipifera,* the *Magnolia acuminata,* the cucumber tree, is the only true member of the magnolia family native to Canadian soils. If it is to be found at all, it will be south of St. Catharines, around Eden and Simcoe and in Ipperwash Provincial Park north of Sarnia. Tracking the north shore of Lake Ontario up as far as Gananoque are a few straggly specimens, all of which are in danger of being cut. This puts *Magnolia acuminata,* the cucumber tree of Canada, on the electric chair of extirpation.

Magnolia acuminata, the cucumber tree, is the largest and the hardiest of the eleven magnolias of the North American continent. In its northern habitat the tree is smaller, multitrunked, with long, sweeping lower branches making an elegant pyramid form as it ages. In its southern habitat, along the rich stream banks of the Smoky Mountains of North Carolina, it reaches its greatest luxury of growth, reaching almost 100 feet (30 m) in 120 years. The cucumber tree naturally consorts with its cousin, *Liriodendron tulipifera,* the tulip tree; *Quercus alba,* the white oak; *Fraxinus americana,* the white ash; *Acer saccharum,* the sugar maple; and various members of the *Carya* or hickory tribe. It is related to another medical tree across the world from China called *Illicium vernum,* the star anise, whose spice is now universally popular.

In the seventeenth century a professor of botany called Pierre Magnol became fascinated with the *Magnolia acuminata* as a tree species. The tree was subsequently named in his honor, *Magnolia,* after his surname, Magnol, and *acuminata* refers to the very sharp point that distinguishes the large boatlike leaves. Of course the aboriginal peoples were well aware of this tree before the arrival of Professor Magnol. They had used it for medicines, for inland canoes, and as a sacred wood for the carving of masks for ceremonial purposes.

M. acuminata has had an important spiritual meaning to many aboriginal peoples such as the Seneca. They carved masks from its wood. These masks had a vital ritualistic purpose in the practice of their daily lives and deep meanings in relation to their interconnected way of thinking.

True to form, wood from *M. acuminata* is soft, heavy, and close grained. It was used for the building of canoes by the aboriginal peoples, the overall quality of the timber being excellent for flotation. So the circle closed when the magnolia canoes were used for transportation around the waterways and rivers where these trees grow so well.

ORGANIC CARE

Magnolia acuminata, the cucumber tree, can be propagated by seed or graft. Because of the fragile status of this tree, seed stocks should be propagated from the most northerly and thus the hardier trees around Lake Ontario. These seeds will be successful in both the northern, zone 4, and southern habitats of

Arisaema triphyllum, jack-in-the-pulpit, a soul companion to *Magnolia acuminata,* cucumber tree. *Arisaema triphyllum* graced the tables of the aboriginal peoples for millennia.

Canada. On the other hand, grafted trees as they age and bear are subject to a variety of internal stresses that sometimes result in rejection or incompatibility of the scion tree with the receiving mother root. For this reason any serious attempt to propagate *M. acuminata* should be done by means of seed.

M. acuminata, the cucumber tree, produces fruit in the fall that has the shape of a squat pickling cucumber. It is about 2–3 inches (5–8 cm) in length. After fertilization, during the summer months, the fruit is green and quite invisible among the foliage. As the days shorten, it becomes pink, turning into a reddish purple at maturity. On drying, it splits open, shedding several bright scarlet seeds that are dropped on white threads to dangle as an "open house." Each seed is about the size of a kernel of corn. It is rounded and flat on two sides. These seeds are ripe in October and can be collected either from the tree or the ground into the end of November. They can be stored for two to three years without any viability change. After collection they should be stored, dry, wrapped in brown paper and placed inside a sealed jar. This is placed in a refrigerator and kept at 32–41°F (0–5°C).

The seeds of the *M. acuminata,* cucumber tree, have dormant embryos. This dormancy must be broken before the seeds will germinate. This can be done by a cold treatment or by a flush of nitrogen-rich water from a thunderstorm in the late spring. Seeds are potted up, ¼ inch (.6 cm) deep. They are maintained at 41°F (5°C) for at least 60 days. In the home, the pots can be sealed in plastic bags and placed in the crisper of the refrigerator for this period. In the warmer growing zones of 7–10, the pots are plunged into a nursery row in April. In zones 5–6, they are placed outside in May, and in zone 4 they are put into sheltered locations around the beginning of June. In all growing zones, the seeds will begin germination as soon as the ambient air is above 75°F (24°C).

Adaptation is the byline of survival on the North American continent. Some species depend on fire for growth, others depend on smoke, and many depend on injury. The seeds of *M. acuminata* have an interior clock that responds to early summer thunderstorms. The rain following thunderstorms, which is rich in nitrogen taken from the air in the form of nitric or nitrous acids, is an immediate trigger to the dormant seed to wake up. This form of nitrogen dependence is not shared by many species.

The seeds germinate very rapidly, producing a 2-inch (5 cm) seedling within five days. The growth is epigeal. The young seedling is very brittle. It has two cotyledons that produce in short order a single acuminate leaf looking a little lopsided. The young growth can be easily damaged because it is so fragile. This first year's growth should have some mechanical protection. It also appears to be very attractive to rabbits and mice, who will seek out these seedlings, ignoring others.

Because of the vagaries of spring freezes in zones 4 and 5, young seedlings can be placed in a cold frame over winter or can be brought indoors into a cool, dark area with temperatures around 35–41°F (2–5°C) for the duration of the winter months. These pots can be moved outdoors any time after May 15 or after the local date of the last killing frost.

M. acuminata, the cucumber tree, likes an acidic soil with a pH in the range of 4–7. Unlike the other magnolia species in North America, it can tolerate some calcium in the soil. Around the neutral pH of 7, it shows some leaf chlorosis or yellowing. Soils with a pH of 6.5–7.0 should have the planting holes prepared in advance of planting, allowing the soil mixture to compost. Then after planting, an acidic mulch of 4 inches (10 cm) of bark or wood chips will help continue the soil surface acidification process.

The planting hole should be prepared with a good quantity of well-aged horse manure. This manure can have a liberal amount of wood chips. Bonemeal (1 pt, .5 l), in the form of steamed bonemeal, or rock phosphate (1 qt, .9 l), should be mixed well into the soil. Peat moss can also be added to the soil. A piece of black plastic is anchored over the prepared hole with stones and left for three weeks to allow the contents to compost. The *M. acuminata* seedlings can then be planted and well watered.

In the colder zones 4–5, some thought should be given to the planting site. A southern exposure with the protection of a building to the north will ensure success. For all zones, wind protection is beneficial because the large leaves of this tree are easily torn and battered. Planting an *M. acuminata* near a tree with large fruit or nuts that drop in the fall should also be avoided for reasons of leaf damage.

As an older *M. acuminata* whip becomes established after transplanting, the leaves will show some yellowing during the first growing season. These will be replaced by greener and healthier foliage in the second and subsequent seasons of growth.

Generally speaking, *M. acuminata,* the cucumber tree, is a very healthy tree, but in rich damp sites without any air movement they may become infested with magnolia scale. Young trees or underdeveloped leaves are usually attacked by *Neolecanium comuparvum,* this particular scale. It appears late in the growing season, usually in August, quite often going unnoticed and overwintering on the tree. In the following spring the poor growth will be noticed. The scale may be easily identified by running a thumb down a young shoot. A spring dormant oil spray will control this disease, spraying on a sunny, dry day with temperatures above or at 50°F (10°C). In zones 8–10, preventative dormant oil sprays can be done in February if necessary.

MEDICINE

The bark, the glandular leaves, the flowers, the fruiting bodies, and the seeds carry the medicine of *Magnolia acuminata,* the cucumber tree.

Over time *Magnolias* have been used to treat a number of minor complaints such as stomach spasm, diarrhea, cramps, and vomiting. But in the warmer and wetter habitat of the South, *M. acuminata* has been used as a substitute for quinine in the treatment of endemic malaria. A quininelike molecule occurs in all magnolias as a double chemical called magnoline. An enzyme chops this biochemical into two molecules that are both active against malaria. The magnolia has another antimalarial component called magnoflorine. The Canadian *M. acuminata* may have further drugs.

A young seedling of *Magnolia acuminata,* an endangered species, taking its first baby steps. It will be kept well watered during its first two years and well mulched.

Hello Dolly

A few years ago I was giving a lecture in southern Ontario. My interest at that time was saving the few remaining cucumber trees that existed. After the lecture a man approached me dangling a huge leaf by the petiole. He smiled, "I had the chainsaw out to this one this afternoon but decided to wait until I heard you speak." I had a cup of tea with my audience as I answered questions. I had in the meantime sent the gentleman off to his house to bring back the fruit. And, yes, it was *Magnolia acuminata,* the cucumber tree, saved from the slice by an hour's wait.

Some things are just meant to be!

The Cherokees used a magnolia bark decoction as a tonic and also as an antimalarial treatment. Two rounded teaspoons of finely shaved early summer bark was added to 1 pint (.5 l) of water. This was hard boiled for 20 minutes and strained. Additional water was added to bring the solution back to 1 pint (.5 l). A teaspoon was taken daily.

The Cherokees also used a bark decoction as a surface douche for rheumatism. Four teaspoons of spring bark shavings were boiled in 1 pint (.5 l) of water. The astringent action of this decoction was also used for sores, wounds, and ulcers. In addition, the cones were collected in the fall. They were dried and ground into a powder. A decoction of this was used as a gentle action skin douche.

The Cayuga peoples used *M. acuminata* as an antihelminthic for the treatment of worms. Two teaspoons (10 ml) of *M. acuminata* bark and two teaspoons (10 ml) of *Viburnum lantanoides,* hobble bush, were put into 2 quarts (1.9 l) of water. This was boiled down to 1 pint (.5 l). Two teaspoons (10 ml) were taken at night. Soft foods were eaten only during the day.

M. acuminata was commonly used by the Seneca peoples as an analgesic for toothache. The inside bark was scraped upward from the eastern side of the tree. This was softened in water until it was moist. The bark was then chewed dry between the teeth to relieve the pain.

The magnolia family has been found by biochemists to be one of just six vascular families with active chemicals against cancer. The active component is thought to be a biochemical called parthenolide, which is also found in feverfew, *Chrysanthemum parthenium.* It is a volatile sesquiterpene lactone found in the flower's fragrance.

The occasional person suffers from a form of contact dermatitis handling the magnolia family. The agent causing this is parthenin. Gloves should be worn.

ECOFUNCTION

The fruits of *Magnolia acuminata* with their enticing, dangling, red seeds attract and feed many seed-eating birds in the fall and early winter. In good years the tree is a prolific producer of seeds rich in both high-quality protein and oil. These are ideal high-energy foods for migratory bird populations.

Because *M. acuminata* is a species that has developed in riparian habitats, the seed design is uniquely adapted for seed dispersal by water. Each seed acts as a small inner tube floating on water. The flat top and bottom makes it move with the smallest current. The circular shape enables it to spin, rotate, and move with the least flow away from the roots of the mother tree, to be trapped by the eddies around the leaves of aquatic plants like water irises. Here they are driven like a bumper car toward wet soil, flat side up, in the perfect position to germinate. The flow dynamics of *M. acuminata*'s red seeds in water is identical to that of red blood cells in the circulatory system of man: both having a flat top and bottom, both circular, both designed for fast and slow fluid dynamics, one to germinate, the other to deliver oxygen, both patterned to be dependent on viscosity, one for blood, the other for mud.

The fragrant flowers of *M. acuminata* produce lactone aerosols that are anticarcinogenic, antimalarial, and also thought to be antiarrhythmic. These volatile biochemicals act as pacemakers for the beating heart. Such fragrances give additional cardiac stamina for migratory birds and mammals who are under the stresses of movement of long distances, especially over large bodies of water from one riparian area to another.

M. acuminata is the host plant for the giant palamedes swallowtail butterfly, *Pterourus palamedes*. This butterfly is attracted by the biofluorescence of the flower of this magnolia, completing its pollination. Any attempt to increase *M. acuminata* will also increase the populations of this butterfly.

The open flower form of *M. acuminata* is used by honeybees, wasps, bumblebees, and other flying insects for both pollen and nectar. These insects are conspicuously absent from this species in European gardens for some reason.

BIOPLAN

Magnolia acuminata, the cucumber tree, should be planted around institutions such as hospitals, trauma and physiotherapy clinics, and retirement homes. The benefits of the fragrance of the flowers would be reaped by the patients. The volatile lactone chemicals from the fragrance go into the airways around the buildings. These lactones give the patients some protection against cancers and malaria and help settle the rhythmic pumping action of the heart. The lactones are absorbed by the damp, mucosal membranes lining the nose and sinuses. They dissolve and are then rapidly transported across barrier membranes into the body.

The *Narcissus* 'Rip Van Winkle'. This late-flowering, double, heritage bulb is an ideal, easy design companion to *Magnolia acuminata,* cucumber tree.

Others who benefit from the fragrance therapy of *M. acuminata* are the mentally challenged or people who suffer from depression or poor mental health. Quite often these groups are on medications that can cause cardiac irregularities. The fragrant lactones help to stabilize such effects. The general calming effect of these fragrances on the sympathetic nervous system is also beneficial to these people. It has an effect somewhat similar to the harmonic mantra of the Tibetan holy men.

The holistic effect of many natural systems such as fragrance, sound, and color are very ancient but are not understood because they affect the deep centers of the brain, such as the limbic region. Fragrance and sound are the first impressions of birth. They travel with us throughout life into death.

The loss of *M. acuminata,* the cucumber tree, from its northern habitat in Canada is a loss of tree species diversity. This has a cascading effect like a stone thrown into a still pond. There is a ratio of around forty insects dependent on each plant species. There are many more times that in fungi associated with underground mycorrhiza and surface lichen growth. These numbers

Amphicarpaea bracteata, hog peanut. It has the curiosity of an aerial seed set for birds and an underground, fleshy, pear-shaped fruit that is much loved by foraging hogs.

are again multiplied for soil bacteria and viruses associated with the tree. Any single one of these creatures could, like the common fungus penicillin, hold the keys for survival.

DESIGN

Magnolia acuminata is a vigorous tree. It is rounded when young, moving to a pyramid form as it ages. The bark is different from all other members of the Magnoliaceae family. Not unlike white ash, *Fraxinus americana,* as it matures, it becomes deeply furrowed. The tree is deciduous, but markedly tropical when in leaf. The leaves are long, 10 inches (25 cm), and boat-shaped with sharply pointed tips. Their color contrasts to other foliage, being a light greenish-yellow, and paler and hardy underneath. The midrib is quite prominent and leaves a horseshoe-shaped leaf scar. It has a bright gray-green, furry bud, like a baby's thumb, pointing upward in the center of it.

Green-yellow flowers occur at the ends of twigs. They are wistfully fragrant. The flowers themselves are quite large, 2 inches (5 cm), solitary, and have an upright, chalicelike form. The bulging, orange-colored stamens can be glimpsed through the yellow-green flower in its sideways position. The flower changes after fertilization into a transformation of green to purple-red as it develops into the fruiting stage. The fruit itself is not unlike a baby rattle with dangling red seeds.

A dozen or so cultivars of *M. acuminata* have been developed, based mainly on blonde variations of the leaf color. *M. a.* 'Cordata' and *M. a.* 'Forma Aurea' are both smaller trees and are both tomentose, or hairy. They reach 33 feet (10 m) at maturity. They are also quite resistant to industrial pollution, making them ideal for smaller, colder suburban gardens in zones 4–5. The flowers in both of these cultivars are also more yellow. A further two yellow-flowered cultivars have been developed from *M. a.* 'Cordata' in which the flowers are larger. These are *M. a.* 'Golden Glow', which has more orange-yellow flowers, and *M. a.* 'Miss Honey Bee' with large, pale yellow flowers.

M. acuminata flowers with the late double daffodils in May into June. The dwarf, double, rare Welsh native, *Narcissus jonquilla* 'Rip Van Winkle', is an ideal bulb companion. It forms a ragged, yellow-green, upright chalice that, too, is fragrant. The fragrance of the Welsh daffodil is refreshingly sweet. Both tree and daffodil are hardy in zones 4–9.

The fall, red color of the dangling seeds can be picked up by the moisture-loving, edible Indian turnip, *Arisaema triphyllum,* or the North American native ginseng *Panax quinquefolius.* Both native species can be used in mass plantings around the shade of the cucumber tree. Both will also survive very well being underplanted with the Welsh native bulb. The combined grouping is lazily beautiful and can last without being disturbed for many years.

Magnolia and Cultivars of Merit

SCIENTIFIC NAME	COMMON NAME	ZONES
Magnolia acuminata	Cucumber tree	4–9
M. a. 'Cordata'	Cordate cucumber tree	4–9
M. a. 'Forma Aurea'	Golden cucumber tree	4–9
M. a. 'Golden Glow'	Golden Glow cucumber tree	4–9
M. a. 'Miss Honey Bee'	Miss Honey Bee cucumber tree	4–9

The gray-green, furry bud that is distinctive of *Magnolia acuminata,* cucumber tree. They become more obvious in the early fall.

Ostrya virginiana, hop hornbeam, at least 200 years old

Ostrya virginiana

HOP HORNBEAM

Betulaceae

Zones 3–9

THE GLOBAL GARDEN

Ostrya virginiana is the old man of the Canadian woods. Even when the tree is enjoying an infancy, it appears to be aged. This comes from a strange lopsided pattern of growth, dandruffed bark, mature, horizontal branches reaching out for sunshine with brown, glinting wrists. These emerge from beneath beards of fall strobili whose papery sacks both taste and smell of beer. These eccentricities, combined with a tendency for producing the cavities with a propensity for rearing a wide range of peering, mammalian heads, all complete the grandfather image.

Ostrya virginiana lies within a small group of four species, each able to produce hoplike strobili, or fruits. This, lies, in turn, within the wide umbrella of the Betulaceae or birch family.

In North America the *O. virginiana* is found from southeast Manitoba east to Cape Breton and as far south to the highlands of Mexico and Guatemala. There is another similar species, the Knowlton hop hornbeam, which is found around Grand Canyon National Park in Arizona, in northern Yavapai County of the San Francisco Mountains, in southeast Utah and Texas, and in the Guadalupe and Sacramento Mountains of New Mexico. It is very rare in these areas and getting more so as time goes by.

In England, southern Europe, and western Asia there is another hop hornbeam, *Ostrya carpinifolia.* This species is quite similar to the Canadian grandmaster but somewhat lacking in character. A smaller tree again is to be found in China, Japan, and Korea called the Japanese hop hornbeam, *O. japonica,* bearing a multitude of fruits in the fall.

The Canadian granddaddy, *Ostrya virginiana,* has borne many common names, ironwood, hop hornbeam, and hornbeam. In the United States it is referred to as American hop hornbeam. These are among the more pleasant names borne by this tree. On the darker side, it also carries a great number of common names based on swear words. This curdling of thought is still used in the farming communities of North America for *Ostrya virginiana.* All of these "cuss words" are, in fact, based on the tree's strengths and durability because saws of all kinds, both ancient and modern, quickly become blunt instruments on short contact with its extraordinarily dense wood.

On the other hand, the North American aboriginal people were much fonder of this tree. The diligent Chippewa named it *Ma'nanous',* using it as a medicine, a health tonic, and a fuel akin to coal. Their high esteem for this tree caused it to be cultivated by the Europeans as early as 1690. It was transported to the royal gardens of England, where it resided as a favorite of the late feisty Queen Mother, a keen gardener.

ORGANIC CARE

Ostrya virginiana is rare in many areas of North America because of sloppy logging practices. This delightful tree does not fit into the linear and board-foot thinking of the lumber world. It has been actively sought out and cut down by the Ontario Ministry of Natural Resources in Canada in mistaken efforts to "improve" maple woods, only to inflict death on the fauna, insects, and bird populations relying on this tree for life.

O. virginiana is propagated only by seed. There is no other means of propagating this tree. Luckily, the seeds are easily identified on the tree. They can be seen in the early fall on the tips of branches as light brown strobili that appear identical to hops on a hop vine. The tree begins to bear after 25 years, and so

Ostrya virginiana, hop hornbeam, with seed-bearing strobili in the fall

older trees should be sought out. These trees will have many branches at breast height and can be easily harvested from the ground.

Depending on the moisture levels in the fall, the strobili will change from light green to green-brown to overall beige-brown. This will begin to happen at the end of August for zones 2–3 and at the beginning of September for zones 4–9. The strobili are handpicked at the green-brown stage into a basket. The seed-filled strobili are allowed to cure in direct sunshine for about three days.

Individual seeds, about 1/5 inch (.5 cm) in length, are in fact small nuts called nutlets. They have a hard pericarp as an outer coating making them easy to pop out of their papery, sacklike covering by hand. When examined closely, each strobilus is seen to consist of twenty or so little paper sacks arranged around a central stalk. Each of these sacks will contain a seed. The seeds from the middle of the strobilus will be the most viable and have a higher germination percentage. These brown, hard seeds that look like miniature hot-air balloons can be collected and immediately planted 2 inches (5 cm) apart in a nursery row. They are covered with ¼ inch (.6 cm) of soil. If possible a light, leaf mulch of *Ostrya* is placed on top of the seedbed. The bed is well labeled.

If the seeds are collected at the right time when they are green, they will germinate the following spring, at which time a light shading is appreciated by the delicate seedlings. These seedlings can be identified at the first true leaf stage after the two, rounded, fleshy cotyledons have formed. The true leaves have sharp double serrations on the edges of the alternating egg-shaped leaves. At this stage they are almost identical to elm, *Ulmus spp*, or apple, *Malus spp*, seed germinating.

If, for some reason, the seeds are collected a little too late in the season into November or when the strobili may have shattered, then the individual seed sacs must be handpicked from the ground. The sacs may also have decomposed, in which case the individual nutlets may be collected from the soil around the tree.

The seeds have to be subjected to a double stratification to break dormancy. This stratification process involves two distinct treatments, one of warmth followed by cold. The seeds are collected and placed into a plastic bag with some damp sand. They are placed on a window sill in full sun for a period of 60 days. The daytime temperatures should be around 86°F (30°C) and the nighttime temperatures should drop to 68°F (20°C). To attain a hot daytime temperature the plastic bag of seeds can be placed on a piece of black poster paper, which will act as a heat sink. A piece of dark roofing tile can also be used.

This warm treatment is immediately followed by a cold stratification. The plastic bag of seeds is placed into a brown paper bag and put into the crisper of the refrigerator for 140 days, at 40°F (5°C). The calendar is marked and the seeds are either potted up or are planted 2 inches (5 cm) apart and ¼ inch (.6 cm) deep in a well-marked nursery row that is mulched. Even then, these seeds may take two years to germinate, with the germination being approximately 25 percent.

Ostrya virginiana are difficult to transplant as older trees. They should be lifted and transplanted to their final position within the first year or two after germination.

O. virginiana likes a deep, rich forest soil of pH 6–6.5. But the tree also grows well in a lime-based soil or a soil with dolomitic substrata and appears to replicate more easily with the availability of calcium to the roots. The tree is a very forgiving tree. It will also grow on poor, barren soils, provided it has some shade produced by the canopies of higher deciduous trees.

In dry sandy soils, the addition of peat moss to the soil is useful. Aged horse manure mixed with maple leaves or maple leaf compost, mixed into the soil of the planting hole, is ideal for growth. A pint (.5 l) each of steamed bonemeal and woodash is also added. This must be added as an amendment in zones 2–3 to increase potassium and potash needed by the tree for frost resistance. *Ostrya* can be planted in shade or in open sunny situations. This tree can also be planted in areas of high winds.

Ostrya are very healthy trees and are rarely attacked by disease. In poor dry soils, the tree is sometimes susceptible to attacks by cankers caused by *Aleurodiscus* sp., *Nectria* sp., and *Strumella coryneoidea.* These heal over and cause attractive pocketing in the mature tree. In dry forests, root rot can be caused by *Armillaria mellea* and *Clitocybe tabescens,* fungi that leave skeletal tree remains that act as perching and feeding posts in the forest. Occasionally, in hot humid summers, powdery mildew, *Microsphaera ellisii,* rears its head. It may be safely ignored, as it is a direct function of weather on leaves that are tough and are particularly water-impermeable. There will be no serious negative effect on the tree.

All of the Ostrya species can be pleached. To pleach is to lace and to interlace branches to form a living archway. Such techniques were used in French and English gardens of the past centuries.

MEDICINE

Ostrya virginiana has been used throughout the ages in North America as a medicinal tree by the many aboriginal peoples who lived alongside this species. Since this form of medicinal practice is an oral one handed down from generation to generation, it is as strong as its weakest link, living memory. Unfortunately much of the medicine using *Ostrya virginiana* has been lost in the past hundred years, but worse still, many of the existing remedies are incomplete.

The medicine of *Ostrya virginiana* lies in the wood. When a reasonably sized piece of wood is split, *Ostrya* wood is seen to have an outer layer of white wood surrounding a dense center of dark wood. This darker wood contains the medical biochemicals.

Ostrya species have not been analyzed for their medical properties. But since they are a member of the Betulaceae family, they could well be expected to share a similar, but slightly different, biochemistry. This family appears to produce two phases of mixed chemicals, one from a birch tar oil and the other from a steam distillation of this tar. The latter phase would be expected to show a similarity to *Ostrya* chemistry. These chemicals are phenol, cresol, xylenol, juaiacol, creosol, and the topical antiseptic pyrocatechol. The aboriginal medical extracts would be expected to have a variety of empyreumatic resins also in trace amounts.

The Chippewa used very small chips of the heartwood in a decoction that was boiled with the fresh green leaves of the *Thuja occidentalis,* eastern white cedar, as the basic remedy of a cough syrup.

The Onondaga aboriginal peoples had a similar syrup for the treatment of cough and catarrh using 1.5 quarts (1.4 l) of heartwood chips in 1 gallon (3.8 l) water boiled down to 1 quart (.9 l). A glassful was taken four times a day.

The Chippewa peoples also used *Ostrya virginiana* heartwood chips in a steam treatment for stiffness of rheumatic joints. A decoction was made by mixing the green tip heads of club moss, *Lycopodium obscurum,* with twig tips of white spruce and chips from the heartwood of *Ostrya virginiana,* hop hornbeam. The steam from the heated, mixed decoction of all three was trapped and directed into the rheumatic area by a blanket or towel. This was done until the skin showed surface sweating.

The vine called American bittersweet, *Celastrus scandens,* which, along with the *Ostrya virginiana,* hop hornbeam, likes to grow in dry places. They were part of the wildwood forests as far south as New Mexico. The surface of this freshly exposed fruit is being groomed for its chemicals.

Hepatica americana, American liverleaf, a dainty spring flower often found growing near *Ostrya virginiana,* hop hornbeam

In addition, the Chippewa treated hemorrhage of the lung in tuberculosis with the inner bark of *Prunus virginiana,* chokecherry, the root of *Corylus* sp., American hazel, the root of *Quercus alba,* white oak, and heartwood chips of *Ostrya virginiana.* The decoction was drunk after steeping the four ingredients together.

The Onondaga had a treatment for tuberculosis itself. It involved gathering medicines two days before the new moon. This is at a time when the hydrostatic chemical draw in a tree is at its least. Again, with reduction in mind, three chips of bark were taken off the north sides of *Quercus alba,* white oak; *Fagus grandiflora,* American beech; *Carpinus caroliniana,* blue beech; and *Ostrya virginiana.* The bark chips were added to 6 quarts (5.7 l) water, which was boiled down to 3 quarts (2.8 l). The medicine was drunk a cupful at a time beginning the second day before the full moon and continuing until the decoction was finished.

There is also a Mohawk treatment of rectal cancer. Bark chips are taken from the side of the *Ostrya virginiana* on which maximum sun shines. The chips are boiled, the medicine drunk.

ECOFUNCTION

Ostrya virginiana has sequential flowering and fruiting in the fall. The seeds or nutlets are consumed by bobwhites, pheasants, grouse, and turkeys. Deer and rabbits browse on the twigs. The new shoots are consumed by porcupines. The male growth tips, catkins, and young shoot tips are eaten by squirrels. Feeding on *Ostrya* goes on into the fall and winter months.

Ostrya virginiana is an understory tree to more mature maples and oaks. This tree acts as an antiturbulent agent in these tall forests. *Ostrya* itself is very strong, tough, and wind resistant. This characteristic is seen even in the leaves, which themselves are particularly tough. This species acts as a wind barrier in the forest, protecting the taller umbrella-shaped canopies from lift. If, in a forest, the *Ostrya virginiana* is removed, the maples and other larger trees will succumb to severe wind damage within a year or two.

The *Ostrya* strobilus is fragrant when crushed, producing an odor that is almost identical to the dried female strobilus of the hop vine, *Humulus lupulus.* This fermenting beer odor is extremely attractive to grouse, who appear to use this fragrance to seek out the *Ostrya* in the fall.

The undersurface of the leaves of *Ostrya* is covered with domatal hairs. These hairs act as protective sites for beneficial insects to take cover from predators. When a caterpillar eats a leaf, the spit from the mouth of the caterpillar interacts with the juice from the broken leaf. This produces an alarm chemical that is a perfume called cis-α-bergamotene, which goes into the airways around the trees. It is an SOS call. The bodyguard insects around respond to this call.

Maples, because of their high sugar content, are prone to fungal infections and diseases of many kinds. It is very probable that the *Ostrya virginiana* leaves, which have active glandular tissue, release their load of topical antiseptics on the ground under the maples and help in the battle against fungal infection for the maple.

As *Ostrya* ages, it becomes a multicavity tree that can house bats, flying squirrels, and mammals of all kinds. It can also become housing to wrens and bluebirds.

Dead *Ostrya* trees have a remarkable ability to remain standing and upright for years in the forest habitat. They become ideal perching sites for predatory birds such as owls.

The inner bark of *Ostrya virginiana,* together with potassium dichromate as a mordant, produces a beautiful orange dye for wool. Alum can be used as a second mordant to produce a lively yellow color.

BIOPLAN

Ostrya virginiana is an ideal tree for city, suburban, or country gardens. It considerably increases the habitat potential of these areas. Its somewhat reduced size and slow growth makes it ideal for small city gardens. Threat to foundations is minimal because of the tree's ability to survive drought. The tree is also naturally tolerant of shade and can be used in north- or northeastern-facing gardens.

Ostrya virginiana should not be planted around institutions such as hospitals or schools. The flood of pollen produced from the male terminal catkins causes rhinitis for many people, especially in zones 6–9.

O. virginiana is extraordinarily wind firm. This tree could be used as a wind buffer in the prairie provinces. The oil industry, in particular, has the need for such wind barriers.

Reforestation projects should use *Ostrya* species in addition to the linear logs for pulping because this tree rebalances the beneficial predaceous harmony within a forest that is nearer to nature and reduces the necessity for spraying.

Ostrya wood has a very fine texture. The growth rings are inconspicuous. The wood is very heavy at 1,350 pounds per cubic yard or 800 kilograms per cubic meter. It takes a remarkable polish and could be used to great effect in the art world.

The wood is also very strong, stiff, and hard, and has an extraordinary resistance to impact. There is no reason why

Allium tricoccum, wild leek or ramp, seeds wait to roll like ball bearings into the fresh damp leaves of a rich woods, to germinate in the spring.

Ostrya wood could not be used for high-traffic flooring in parquet designs with other woods.

DESIGN

Ostrya virginiana usually grows to about 30 feet (9 m) tall with the occasional miscreant growth of 60 feet (18 m). In the garden it is an attractively elegant small tree that holds interest in spring and reappears again in the fall with its load of strobili and rich, warm, yellow-tinted leaves. During the winter months its form and uneven trunk add considerable interest to the landscape. The tree's form stems from the unusual pattern of growth found within the tree. It does not have typical apical meristems. Instead spring growth commences from secondary buds that are set at angles of 45°. The first flush of spring growth is the 2-inch (5 cm) male catkin coming before the leaves.

The alternate leaves are 3 inches (8 cm) long and 1.5 inches (4 cm) wide. They are slightly heart-shaped at the bases with tapering tips and saw-toothed margins. When fully grown the leaves

A simple seat waiting for the summer shade of hop hornbeam, *Ostrya virginiana*

appear to be very thin. The upper surface is a darker yellow-green than the undersurface, which is also hairy.

Plantings of sharp-lobed liverleaf, *Hepatica acutiloba,* with its white, pink, or blue fragile flowers and its low-growing habit or the more dainty American liverleaf, *Hepatica americana,* with its white or pink flower and densely hairy leaves contrast beautifully with the scruffy spring form of *Ostrya.*

The hop vine, *Humulus lupulus,* can be planted near the *Ostrya virginiana* in the garden to multiply the effect of the fall strobili. In August and September the strobili have a waterfall effect on the trees, cascading the eye down to the ground, where a low piece of sculpture, a stone, or a small, squat, water bath can be placed for additional effect.

In warmer gardens, zones 5–9, the rare Mongolian *Ostryopsis davidiana* shrub could be planted, whose long hoplike fruit is an attractive feature in an early winter garden.

Ostrya Species and Cultivar of Merit

SCIENTIFIC NAME	COMMON NAME	ZONES
Ostrya japonica	Japanese hop hornbeam	4–9
O. virginiana	American hop hornbeam	3–9
O. v. 'Knowlton'	Knowlton hop hornbeam	3–9

Ostrya virginiana, hop hornbeam. Some of them show a very attractive exfoliation of the bark as they mature.

Pinus strobus, white pine. With its wedding band of American bittersweet, *Celastrus scandens*, it is a tree of legend. When this tree starts to stumble, so will we. It is now having difficulty regenerating in forests.

Pinus

PINE

Pinaceae

Zones 3–10

A second-cut mixed pine forest. It still shows the integrity of mixed tree species.

THE GLOBAL GARDEN

The birthright of a white pine virgin forest that produced 750,000,000,000 board feet of lumber is no more. The timber barons are dead. The money rides on in the castles of Europe. Nobody has noticed; global warming will clean the slate.

Invisibly trapped in this cycle is art. The now famous images *The Jack Pine* by Tom Thompson and *The White Pine* by A. J. Casson, tell us the story of a hidden history in the heart of North America. It is the tree. The tree, in particular the pine, was and is the true treasure of the aboriginal peoples of this continent. These artists merely trapped the images of what was there before, a magnificence, a diversity of unspeakable glory. And we are letting it go. For who among us can replant that forest? Who among us will preserve the slate?

The pine is first and foremost a tree of medicine. All over the global garden this knowledge is ancient. The pharmacy of the pine is as common in Turkey as it is in the Balkans, as it is to the Chinese, as it was to the ancient Picts of Scotland and now, as it is to us. Hard though it may be to conceive, some pines such as the bristlecone pine, *Pinus aristata,* transcend just two to three generations from the last ice age. Their life span is 4,000 years, spitting out a crop of seeds every 102 years. The bristlecones now just at middle age have seen the birth and death of Christ. They have continued maturing into this millennium, logging an additional 2,000 years of growth under their aging bark. Other pines, such as the sugar pine, *P. lambertiana,* in the Sierra Nevada, will reach 245 feet in 600 years, making them one of the tallest living creatures on this planet. The paradox of these 25-story giants is that water flows into the mesophyll cells at this extraordinary height and science can identify the plumbing but not the force that drives it.

Apart from lumber, many pines produce food in the form of pine nuts. The most familiar are the Italian stone pine, *Pinus pinea,* and the Korean nut pine, *Pinus koraiensis,* which is a native of eastern Russia and northern Japan as well as Korea. Its growing zones are 3–10. The Swiss stone pine, *Pinus cembra,* zones 4–10, takes 30 years to produce its first nut crop. The Mexican pinyon pine, *Pinus cembroides,* may be grown in a large range, zones 3–9, because it grows naturally to an altitude of 7,500 feet (2,300 m). These seeds are large and oily. The tree matures in 350 years, setting seeds in a 2- to 7-year cycle, like all pinyons. The most common pinyon nut to be found in our local grocery store is the Colorado pinyon pine, *Pinus edulis.* The nut is large, oily, both sweet and nutritious. The tree will grow up to

The male flowers of *Pinus resinosa,* red pine. These pollen factories are filled to capacity.

altitudes of 9,000 feet (2,700 m), making it capable of growing in zones 3–9. A mountain dweller, the unique single leaf pinyon pine, *Pinus monophylla,* zones 4–9, has chocolate brown seeds that have fed the aboriginal peoples of Nevada and California throughout the centuries. The Parry pinyon pine, *Pinus quadrifolia,* grows in zones 7–10 in California. Its reddish-brown nuts are highly nutritious. Warmer again is the scrawny Digger pine, *P. sabiniana,* zones 8–10, famous for its huge 10-inch (25 cm) cones, whose edible nuts are as large as broad beans. These were largely collected by the Digger peoples of California in times of drought and scarcity. Commerce and agriculture have yet to exploit these trees for food.

Many pine species, among them *Pinus clausa,* the sand pine, wait for fire to open their cones. Some old field species, *Pinus taeda,* the loblolly, and *P. echinata,* the shortleaf pine, swarm into tired old mineral soils of abandoned farms to reclaim them for greener days. The pocosin pine, *P. serotina,* craftily grows in a pocosin or dry swamp for fire protection.

There is the pitch pine, *P. rigida,* which will handily stool, unlike all other conifers, and can be coppiced. Many pines, such as the red pine, *P. resinosa,* are resistant to fire. The mature trees will not burn, but the cones wait for the heat to open the scales to release seeds. So fire means life for the red pine.

Canada has had its own taste of the nonnative scotch pine, *P. sylvestris,* the most planted and most studied pine in the world. It hails from the eastern Scottish Highlands and south of Galway Bay, Ireland, the last living remnants of the seven-thousand-year-old ancient pine wildwood. Canada boasts nine species of pine, eastern white pine, *P. strobus;* western white pine, *P. monticola;* limber pine, *P. flexilis;* whitebark pine, *P. albicaulis;* ponderosa pine, *P. ponderosa;* pitch pine, *P. rigida;* red pine, *P. resinosa;* jack pine, *P. banksiana;* and the lodgepole pine, *P. contorta.* Then there is the pine made famous by John Steinbeck, the Monterey pine, *Pinus radiata,* growing in its tiny California lair. Associating with it is the wonderful bishop pine, *Pinus muricata.* These pines in turn consort with a party of snobs, the California live oak, the Gowen cypress, Douglas fir, madrone, tan oak, bigleaf maple, California laurel, the ceanothus, and the rhododendron.

ORGANIC CARE

All pine species can be grown from seed, by rooting, by grafting, and by transplanting of pot-grown nursery stock. Tourists with green thumbs who travel to warmer pastures should know that their caches of exotic pine seeds are prone to short dormancy times, fast growth rates, and long growing seasons, only to be followed by damage from the cold temperatures of their winter. Unhappily, southern pines do not travel north very well, but northern pines can trek south, thrive, and grow extremely well in their adoptive land. All north temperate Canadian pines should be planted in the spring. They appreciate a rich garden soil as much as any other plant does, but they are most forgiving and will make good growth on poor, lean soils. Their preference is for a slightly acid soil with a pH of 4–6.5.

If pines are to be propagated by rooting, the tip cuttings should be taken from trees that are not older than five years. Root cuttings and graftings are used by the nursery trade to propagate particularly rare cultivars, which are, in turn, equally rare in price.

All enthusiastic gardeners can grow pines from seed. Indeed pine seeds can be stored in brown paper in sealed glass jars at 32–41°F (0–5°C) for up to 30 years with little reduction in germination.

Male and female cones occur on the same pine. In north temperate areas, cones are collected in late fall and early winter. They should be "the pick of the litter." The mother tree should not be diseased. It should have other pines around to ensure pollination for a greater number of viable seeds. A mature seed has firm white to cream-colored endosperm with a white to yellow embryo that nearly fills the seed cavity. Most pines, except for the edible pine seeds called pinyons in the United States and *piñones* in Spanish, have a papery wing that continues around the embryo as a protective cover. The home gardener should keep this form intact for planting. By contrast, an immature or nonfertilized seed is dun-colored, light in weight, and dry looking.

Pinus koraiensis, Korean pine. Like most pines with edible nuts, it produces large seedlings that must be coddled in their first year.

Plenty of cones can be collected and placed on dry newspapers in a warm, well-ventilated room. The cones open to release their seeds. These are collected and planted in a special soil mix. When the cones are being collected, some soil is removed, about 1 quart (.9 l), from around the base of the mother tree. The top layers of pine needles are moved back to expose some decomposing pine needles that are in direct contact with the soil. This material is mixed with the potting soil as an inoculant. The seeds are planted flat on the soil surface and covered with ½ inch (1 cm) of soil. The young pine will grow to a tiny green feather seedling in 30 days. This is placed into a larger pot the second year, carefully conserving the original, inoculated soil. The young pines can be planted out and allowed to grow on in the third to the fifth year. The inoculated soil from the pot should be included in the planting hole. In times of drought the young pines should receive some water.

There is a disease of five-needled pines that was introduced into North America and has devastating consequences for these

Clematis × durandii, Durand clematis. With its huge, blue flowers lasting from June into October, it can be allowed to climb through a dwarf pine in a careless fashion in the garden.

trees. In Canada, the following pines are affected: eastern white pine, *Pinus strobus;* western white pine, *P. monticola;* limber pine, *P. flexilis;* and the whitebark pine, *P. albicaulis.* The disease, a pine blister rust, is caused by a fungus *Cronartium ribicola,* which cohosts on wild gooseberries, *Ribes hirtellum,* and black currants, *R. nigrum.* Global warming is making this disease more infectious. The *Ribes* need alkaline soil for their wellbeing, but global warming is acidifying all soils in the global garden, including that for the *Ribes.* This is in turn dropping the immune system of these fruiting shrubs and making them open to infection by this rust fungus.

The fungus is quite easily recognized as bright fluorescent orange pustules on the leaves and fruit of the *Ribes.* These heavy spores, called ascospores, are blown toward the five-needled pines and infect them. The pine takes several years to die. A barrier distance of 900 feet (275 m) between the *Ribes* and the pine will help to control this fungus. One black currant, *R. nigrum* 'Consort', is immune, as are the red, white, and western *R. odoratum* species and many other *Ribes* cultivars.

MEDICINE

Pines all over the world are living pharmacopoeias. Many pines exist in extreme conditions of drought, heat, and elevation. Like most medicinal herbs, these naturally stressed, tough survivors will have greater medicine. Chemical isomeric compounds from the pine in North America are dextrorotatory. Chemicals from the same pine species in Europe are levorotatory. Rotation of ordinary light through a solution of a pure chemical is a means of characterizing that chemical for a chemist. It is a way of finding its family name. Isomers in medicine that are *d-*, or dextrorotatory, and *l-*, or levorotatory, are not all used in equal proportion by the human body, which seems to want the *l-* or levorotation product. This isomeric difference is not fully known, not understood, and not thoroughly investigated. It may have great importance in medicine.

Historically, pines are the source of many essential oils and substances such as pine oil, pine needle oil, pine tar, and turpentine. Pine is also the source of pinene, which is used in the manufacture of camphor, insecticides, solvents, plasticizers, perfume bases, and synthetic pine oil. Pine is also the source of an important antibiotic and fungicide called pinosylvin, *trans-*3,5-dihydroxystilbene, which has undoubtedly been the cause of abortion in farm animals eating an excess of pine leaves and branches.

North America is rich in pines, and consequently the aboriginal peoples possess many cures and uses for this tree. The Oneida peoples had a clever means of fumigation as a disease preventative. Fresh pine needles of *Pinus strobus,* white pine, were smoldered in a pail. The smoking pail was brought into the house. The house was sealed. This was done twice a year in spring and fall. This was an antiseptic treatment that included sterilizing even the air. Modern methods using pine products are not as effective as the older fumigation method against airborne disease.

The value of pine knots have been known for thousands of years on both sides of the Atlantic. The Romans used them as night flares. These night lights also had antiseptic and antibiotic action while burning. The Cayuga peoples, on the other hand, used these knots in the treatment of tuberculosis. They consider them to be superior to penicillin.

Generally speaking, pine knots have the highest content of the antibiotic pinosylvin. The Cayuga chose the most potent of these knots and saved them for medicine. These came from pinewood where the heartwood had completely decayed around the knots. A total of twelve knots 14 inches (35 cm) long were taken and split lengthwise to expose the central pith tissue. These were boiled in 6 quarts (5.7 l) of well water that was reduced to 3 quarts (2.8 l). A dose of 4 ounces (125 ml) was taken three times a day after meals on the first day. On the second day 3 teaspoons (15 ml) three times a day was taken after meals. This treatment is unique because of the means of selection of the pine knots and the method of opening them at the pith core, releasing the highest possible content of pinosylvin antibiotic. It is the main antibiotic cache of the pine tree.

Throughout the ages the most popular use for pines was in the treatment of respiratory diseases. This included colds, coughs, laryngitis, chronic bronchitis, catarrh, sore throats, and asthma. This is because the pine exerts a dilatatory or opening action on the bronchi of the lungs. Since ancient times even a walk through a mature pine woods in summer was considered to be beneficial to one's general health. The fresh leaves exert a stimulant effect on breathing with the addition of mild anesthetic properties. There is possibly some mild narcotic function also in pines. In warm air the pine sweats a natural monomethyl and dimethylester of pinosylvin, which are both aerosols.

In the wintertime, the Onondaga peoples used pine bark and needles to make a steam that was inhaled. This was followed by making a salve from pine resin that was heated with beeswax and animal tallow to which honey was added. A medicinal lollypop was made of this and the patient, usually a child, sucked his or her way to health again. This salve was kept in winter storage for the treatment of open cuts and wounds.

Modern baby powder, when dusted on bottoms, simply absorbs moisture. But aboriginal peoples used a baby powder from pine wood. This powder was antiseptic as well as absorptive. Well-rotted pine wood was ground to a fine dust. It was carefully stored and used by mothers, especially if the baby showed chafing with teething. This powder has a mild fungicidal and antiseptic action on young skin. It was also used for umbilical cord healing much as alcohol swabs are used today.

On the darker side of life, *Pinus strobus,* the white pine, was used as a ghost medicine. It was used as an emetic when someone died and the death could not be forgotten. It is used to wash the eyes of someone who had seen a dead person. Windblown branches were burned, and the smoke was left to filter into the eyes. It was used when a household member had returned after a long time. Pine smoke was allowed to go through the house. These ghost medicines were practiced by the Seneca peoples.

Pine knot salve mixture, the older the better, was used to treat poison ivy and insect bites. It was also used in the treatment of menopausal males when skin cracking of the penis was a problem. This situation is exacerbated in modern living by the presence of pesticide xenoestrogens in the environment. The salve, a mixture of *P. strobus,* white pine, and *Aralia nudicaulus,* wild sarsaparilla, with tallow and beeswax was applied daily as a protective, antiseptic barrier skin cream.

The burning of needles of *Pinus rigida,* pitch pine, was used to rid an area of fleas, and the sweetened pitch was used to treat rheumatism.

ECOFUNCTION

The pines in the global garden north of the equator function as air fresheners. This is very evident when the temperatures are elevated and the humidity levels are high. The pines release terpenes and many other volatile, hydrophobic compounds into the atmosphere. These include turpentine; the quaiacol group, most of which are expectorant, antitussive, anti-inflammatory, antiulcerative, and muscle relaxant; creosol; methylcreosol; phenol; phloral; toluene; xylene; and other unknown hydrocarbons. All of these compounds in the amounts and ratios emitted from pines are beneficial to man, animals, and the flying insect world.

Propolis is collected from pines, among other species. The honeybee can be seen tearing at the resin with her mandibles. It takes her a quarter of an hour to load this propolis. When she arrives back at the hive, other worker bees help to unload. This is why spots of propolis are found around hive entrances. Bees mix propolis with wax and this becomes a fungicidal, antiseptic, and antibiotic wallpapering for the inside of the beehive. The fact that propolis is an old folk medicine in northern European forested areas is not surprising.

Most people know that honeybees make honey and the raw honey is nectar collected from flowers. Few people also know that bees also collect propolis. Propolis is sometimes called bee gum. Incidentally, it is an ingredient in the varnish that gives the

Clematis campanula (syn. *C. campaniflora*), campanulate clematis. It can be allowed to grow through pines in a careless fashion.

The bristlecone pine, *Pinus aristata,* as a mere toddler of 35 years. It still crawls on the ground awaiting its 4,000-year future.

Stradivarius violin its tone. It varies in color and in character. It is used to stop up cracks and crevices in the hive. It is also used to change the entrance of the hive in times when the bees' population is reduced due to many domestic matters in the hive such as illness, queen renewal, or weather change. There is also an ethnic aspect for this; some bee races use more propolis than others.

The pine is associated with at least 53 different fungi, many of which are edible, making pine woods a feast for the fall. The mycorrhizal association of these fungi with the pine makes the habitat of the pine unique and is the reason why so many foreign species cannot be established without soil inoculation. This means that underground, the pine associates in a special way with an extraordinary diversity of fungal life, the essence of which is not understood.

Seeds from pine cones are food for many songbirds such as the pine siskin, the pine grosbeak, and the crossbills. Because pine seeds are naturally shed throughout the winter months, the pine seeds are also food to many mammals in barren areas. The wingless seeds known as pine nuts, especially the larger nuts such as *Pinus lambertiana,* the sugar pine; *P. torreyana,* the Torrey pine; *P. sabiniana,* the Digger pine; and *P. quadrifolia,* the Parry pinyon, have had considerable historical importance as famine foods for the aboriginal peoples throughout the ages in North America. Pine needles are also host to the eastern and western pine elfin butterflies.

BIOPLAN

The pine is not a city dweller, preferring, towns, villages, and backwater places. It is as an old growth pine forest that the true majesty of the pine is experienced. There are a few meager examples of pine virgin forest left in Canada. But there are some second-growth forested areas where the rushing winds are amplified through the needles as sounds like the running of water, where the undergrowth is a soft carpet of needles underfoot, and where the shortening days of summer fire the floor with a colorful stippling of mushrooms of extraordinary diversity, a sight that is long remembered and few have seen.

Pines should be part of a bioplan around any hospital setting. Space should be allocated so that fifty to one hundred trees of *Pinus strobus,* the eastern white pine, or *P. rigida,* the pitch pine, can mature in one block. This area should remain undisturbed, and the pine needles should accumulate into carpeting. A boardwalk could meander into the heart of this small forest for sitting and relaxation. This space would help lung transplant patients, chronic asthmatics, tuberculosis patients, and people with respiratory diseases of allergenic origin, which are climbing in numbers worldwide. Air flow from these pines could be selectively vented into the hospital itself.

Since colds, flus, and coughs are the mainstay of childhood complaints, pines should be part of the bioplan for nursery, kindergarten, and junior schools. Such plantings improve health in crowded conditions. The air would have a natural antiseptic that would help to naturally control diseases of viral and bacterial origin.

Pines should be bioplanned into areas around homes for the aged and the chronically ill in such a way that both the aerosols are effective and the sounds of the wind in the boughs are heard. The harmonics of the latter have a soothing effect on the ill.

Historically, pine forest retreats have been used in the past as health spas. Canada and Europe show the ruins of many of these resorts. In the Americas they often followed the railways from the Atlantic to the Pacific. World health has not improved by their demise.

If the firecracker *Pinus roxburgii,* the emodi pine, which is loaded with turpentine, can be used for Chinese fireworks in China, then *P. rigida* can be looked at in the same way. *P. rigida,* the pitch pine, is a North American native reduced to a mere handful of trees because this species was used for "naval stores" for creosote, pitch, and turpentine a century ago. It, unlike its emodi cousin, will regenerate from the stump and can be coppiced. This species will also grow in barren land, being an ideal species for biomass energy regeneration by burning. It can be considered to be a sustainable energy crop producing "green power" into a hydroelectric grid system. The carbon banking of its growth produces a low flash point wood ideal for set-aside biomass production. The banking of carbon during its growth cancels out the carbon dioxide production from burning for power.

The pulp and paper business should seriously examine pine coppicing as a means of sustainable pulping for paper production. A possible mixture of *Cannabis sativum,* gallow grass, with *Pinus rigida,* pitch pine, pulp would produce long-staple fibers that could be recycled into an ever improving paper quality with the same starting product. Present-day paper cannot be properly recycled due to the fragility of the paper fibers in the second round of treatment for paper making.

The North American continent has the greatest number of pine nut–producing pines. These have never been exploited as a food source, as they have been in Italy, Russia, and China. The nut meats for fresh eating and for confectionary are highly desirable.

A high-quality oil can be readily expressed from these nuts. A bioplan for a nut orchard should include some or all of these North American natives for cultivation and production. These are *Pinus albicaulis,* the white bark pine; *P. cembroides,* the Mexican nut pine; *P. coulteri,* the coulter pine; *P. edulis,* the Rocky Mountain pine; *P. flexilis,* the Limber pine; *P. lambertiana,* the sugar pine; *P. monophylla,* the single leaf pinyon; *P. monticola,* the western white pine; *P. ponderosa* var. *arizonica,* the Arizona pine; *P. ponderosa* var. *ponderosa,* the ponderosa pine; *P. ponderosa* var. *scopulorum,* the Rocky Mountain ponderosa pine; *P. quadrifolia,* Parry pinyon; *P. sabiniana,* the Digger pine; *P. strobiformis,* the southwestern white pine; and *P. torreyana,* the Torrey pine. There is also *P. aristata,* the bristlecone pine, that produces a reasonably edible seed that can vary in size depending on the growing conditions of that particular century. There is also the very rare *P. washoensis,* the Washoe pine with its beautiful black cones. This species deserves to be brought back from obscurity.

DESIGN

Pines come in all shapes and sizes. Many of the popular North American species have garden cultivars. The most commonly planted pine ornamentals are the Austrian pine, mugho pine, Scotch pine, and the white pine, which is popularly called the Weymouth pine in Europe after Lord Weymouth, who brought it to England in the 1700s. This large pine also has some interesting garden cultivars: *Pinus strobus* 'Compacta' is a dwarf form with a low habit, and *P. s.* 'Fastigiata' has a wonderful columnar habit, ideal for a limited space. *P. s.* 'Pendula' has a graceful weeping habit, so useful in a cold, northern garden. The standard with blue-green needles is *P. s.* 'Uconn'. The imported Scotch pine, *P. sylvestris,* also has a dwarf form, *P. s.* 'Nana' or *P. s.* 'Watereri'. The arolla pine, *P. cembra,* which is also known in Canada as the Swiss stone pine, has some pleasing cultivars such as *P. c .* 'Jermyns', a very slow growing dwarf form. There

Pinus rigida, pitch pine, one of the handful remaining on the shores of the St. Lawrence River near Rockport, Ontario. They were raped for their seafaring chemicals in the 1800s.

Pinus Species and Cultivars of Merit

SCIENTIFIC NAME	COMMON NAME	ZONES
Pinus albicaulus 'Algonquin Pillar'	Whitebark pine	4–10
P. aristata	Bristlecone pine	2–10
P. banksiana	Jack pine	2–10
P. bungeana	Lace bark pine	4–10
P. cembra 'Pygmaea'	Arolla pine	3–10
P. c. 'Jermyns'	Jermyns arolla pine	3–10
P. c. 'Aureovariegata'	Variegated arolla pine	3–10
P. gerardiana	Gerard's pine	4–10
P. mugo 'Gnom'	Gnom mountain pine	2–10
P. strobus 'Compacta'	Compact white pine	2–10
P. s. 'Fastigiata'	Fastigiate white pine	2–10
P. s. 'Pendula'	Weeping white pine	2–10
P. s. 'Uconn'	Uconn white pine	2–10
P. sylvestris 'Nana' (syn. *P. s.* 'Watereri')	Dwarf Scots pine	3–10

is also a yellow-leafed cultivar, *P. c.* 'Aureovariegata'. Again the bristlecone pine, *P. aristata,* is becoming popular, as is the pyramidal *P. albicaulus.* Often overlooked for its ornamental potential is *P. banksiana,* the common jack pine, which, when standing alone, assumes an airy, irregular beauty in the garden.

Rare, but becoming more known to the gardening community, are the two lace bark pines, *P. gerardiana,* Gerard's pine, and *P. bungeana,* whose spectacular trunk patchwork of white, yellow, purple, brown, and green coloring elicits sighing admiration.

If pines are used alone in a small garden, they can be grouped for visual interest and can, in turn, be mixed with ornamental grasses such as *Festuca glauca, Pennisetum alopecuroides,* or *Helictotrichon sempervirens.*

In a larger city garden or by a gate entrance a *Clematis,* such as *C. campanula* with tiny blue campanulate flowers or *C. rubrotrimarginata* with fragrant red autumnal flowers or *C.* × *durandii* with a full season's show of startlingly large blue flowers can be allowed to grow through the pine in a careless fashion. The deliberate combination of textures gives a richness to any garden.

North American Edible Pine Nuts of Merit

SCIENTIFIC NAME	COMMON NAME	ZONES
Pinus albicaulis	Whitebark pine	3–10
P. aristata	Bristlecone pine	2–9
P. cembroides	Mexican nut pine	3–9
P. coulteri	Coulter pine	4–9
P. edulis	Rocky Mountain pine	3–9
P. flexilis	Limber pine	4–9
P. lambertiana	Sugar pine	4–9
P. monophylla	Single leaf pinyon pine	4–9
P. monticola	Western white pine	4–9
P. ponderosa var. *arizonica*	Arizona pine	4–9
P. p. var. *ponderosa*	Ponderosa pine	5–9
P. p. var. *scopulorum*	Rocky Mountain ponderosa pine	4–9
P. quadrifolia	Parry pinyon pine	6–9
P. sabiniana	Digger pine	5–9
P. strobiformis	Southwestern white pine	5–9
P. torreyana	Torrey pine	6–9
P. washoensis	Washoe pine	5–9

An elegant statement of growth. A *Pinus strobus,* white pine, cuts into the sky with its typical silhouette.

Ptelea trifoliata, wafer ash. In a tiny garden, it will dominate.

Ptelea trifoliata

WAFER ASH

Rutaceae Zones 3–9

THE GLOBAL GARDEN

The Cinderella of the Canadian forests is the wafer ash, *Ptelea trifoliata.* This small tree sits with the great unwashed. It is unnoticed, unloved, and unwanted. And like Cinderella, this tree holds a magic spell that can change humanity forever.

In another time and in another place in North America, when the natural knowledge systems of the aboriginal peoples held sway, the wafer ash, *P. trifoliata,* was known. This was in an age when wisdom in its accumulated and amplified form came from the dreams of tribal medicine men and women of the Menominee and Meskwaki nations. The medicines of the wafer ash, *P. trifoliata,* were extraordinary. They used this tree as a herbal panacea. They called this tree the Sacred Tree.

The botanical story is fascinating because the wafer ash, *P. trifoliata,* is the "one who got away." In North America the last ice age was a crushing blow to the heat-loving tropical plant species on this continent. As a tropical tree, the wafer ash, *P. trifoliata,* shares its family fold with the citrus species, oranges, lemons, grapefruit, and mandarins. These are all members of the famous rue, or Rutaceae, family. The wafer ash learned how to survive the cold in a genetic dance.

Botanical reports from Mexico to California to Ontario indicate that the tree varies its leaf to survive. In Mexico, *P. baldwinii* has leaves that are smaller and are reflective. In California the leaves of *P. trifoliata* var. *crenulata* have a more spongy mesophyll for water storage. As the plant travels north, the cuticle protecting the leaves of *P. nitens* is thicker and there is more chlorophyll in it, trying to take better advantage of a decreasing sun.

Chance has also helped the wafer ash, *P. trifoliata,* to survive. This tree's chosen habitat is rocky and barren. These kinds of living conditions produce the warmth of radiant energy from the "black box" effect of the surrounding rocks during the cold winter months. The seeds, too, are a bit strange. Each one of them is shaped like a perfect flying saucer with light reticulation like a spider's web around the central heavy embryo. This helps the seed in dispersal. There is also the sheer, whimsical idea that Cinderella, as part of the ancient natural pharmacopoeia, might have been saved and planted by those who lived by an oral history. The answers to this we will never know.

Ptelea trifoliata (syn. *P. isophylla),* goes by the common names of wafer ash, hop tree, wooly common hop tree, shrubby trefoil, and stinking ash. Wafer ash is a small tree with a graceful rounded crown spreading as it ages. At maturity it is about 20 feet (6 m) tall. In the late spring, the tree is covered with corymbs, or clusters, of small cream flowers that are the most fragrant flowers of any hardy tree in the north temperate forest. The fragrance is strongly honeysuckle-like, more powerful and sweet than the wild woodbine, *Lonicera periclymenum,* of Europe and Morocco. The flowers are followed by hanging clusters of oval, winged fruits somewhat like the elm, *Ulmus.* These fruits change color from dense green to brown as they persist on the tree during the winter months. The bark has a delicate appearance, smooth and striated similar to other members of the citrus family. The elegant, trifoliate leaves have a long leaf stalk. These, depending on the cultivar, may be a soft yellow or a dense forest green. But all parts of the wafer ash, *P. trifoliata,* the leaves, the fruits, the petioles, the bark, and even the twigs, produce a strong fragrance of orange when crushed or injured in any way.

Ptelea trifoliata, wafer ash, with an unmistakable beauty in flower bud

ORGANIC CARE

The wafer ash, *Ptelea trifoliata,* can be propagated by layering, grafting, budding, and by seed. The wafer ash grows readily from seed. The seeds, or samaras, are collected in October when they are white-green in color. The seed is air dried on newspapers at 70°F (21°C) until the seeds have turned a dull cream brown. When the seeds are completely dry, a reticulate network opens up around the embryo and it becomes both mature and dormant. At this stage of development the seeds can be planted outside in a nursery row ¼ inch (.6 cm) deep and 2 inches (5 cm) apart.

The dormant seeds can also be stratified indoors to break dormancy. They are placed into dampish sand in a plastic bag for three months at 40°F (4°C) in the crisper of a refrigerator. After this time the seeds are potted up and are subjected to daytime temperatures of 65°F (18°C). This is dropped to nighttime temperatures of 50°F (10°C). The cycle is maintained until the seeds germinate. The seeds are moved into the direct light. In the colder zones of 2–5, the young seedlings, which are quite frost hardy, can be placed outside after May 15. They should have a cover of Reemay over them until June. Usually the seeds take about twenty days to germinate. They are immediately frost hardy and very drought resistant. The seedlings grow very rapidly to two interrodes worth of growth and fill out for the remainder of the season. The seeds that have not germinated in the first year will do so the following spring. At any time these seedlings can be transplanted. If they are watered well one hour prior to transplanting, they appear to have little or no transplanting shock. This is true for two- and three-year-old seedling trees.

The wafer ash, *P. trifoliata,* like all of the *Ptelea* species and cultivars, prefers full sun conditions. It will also grow well in light shade. It likes a rich fertile soil but will also grow resoundingly well in a poor barren soil provided the soil has a supply of water. The quality of the soil itself is not important for the wafer ash. It must be porous with excellent drainage in summer and especially in winter. Like all of the citrus family, the roots seem to appreciate a high level of oxygen in the soil. This can be easily achieved by the addition of some gravel at the base of the planting hole. Generally speaking, the addition of calcium in the form of dolomitic lime or small marble chips seems to be appreciated by this tree, which appears to prefer a neutral soil, around a pH of 7.0.

Throughout its life into maturity the wafer ash, *P. trifoliata,* is the most trouble-free of all of the smaller trees. It is not affected by insect damage, nor is it grazed upon by the larger mammals like deer or rabbits. It definitely earns its keep with its quiet, green, steadfast beauty into the fall.

MEDICINE

The medicine of the wafer ash, *P. trifoliata,* is found in the leaves and fruit, in the young shoots and branches and in both trunk and root barks.

If a mature leaf of wafer ash, *P. trifoliata,* is held up to the light, a series of tiny dots are seen on the underside of the leaf. These dots are found all over the tree, on twigs, branches, and bark. These are secretory glands carrying most of the medicines of the tree. These glands contain volatile oils, limonene, various quinols, resins, tannins, coumarin, furocoumarins, the rhamnoglucoside, rutin, byangelicin, marmesin, marmesinen, and other alkaloids, steroids, choline, the tropane alkaloid, scopolin, skimmianine, sugars, and many other active substances.

The Menominee aboriginal peoples used the bark of the wafer ash, *P. trifoliata,* as a general health panacea. They found that the bark had the capacity to synergize other medicines. In other words, the medicine from *P. trifoliata* could piggyback on other chemicals and make them more effective as medicine.

From around the world there are only a handful of medicines that are considered to improve the health of the human body functioning as a whole. These are the ginseng group, *Eleutherococcus senticolus,* Siberian ginseng; *Panax ginseng,* oriental gin-

Flowers of *Ptelea trifoliata,* wafer ash. A wonderful fragrance of honeysuckle is the dominant scent given off.

seng; and P. *quinquefolius*, or American ginseng. Again, from North America there is *Eupatorium perfoliatum*, boneset. There is a Sudanese plant from the Nuba mountains in Africa called *Cochlospermum niloticum*, whose swollen root nodules are used. There is a lily in the Asian tropics called *Hypoxis aurea*. Back to North America there is *Coptis trifolia*, golden thread, which was used in a dry form like its famous cousin C. *teeta* in Egypt. There are a few more in the tropics, and there is wafer ash, P. *trifoliata*.

The medicines of the wafer ash, P. *trifoliata*, are very strong and should only be used by informed medical personnel. Because of its high coumarin content the tree should not even be handled by pregnant or by lactating women.

The medicines have been used to increase appetite because it increases blood flow to the gastrointestinal tract. It has been used as an antibiotic and in the management of vitiligo, psoriasis, and mycosis fungoides. It is an antihelminthic for both roundworm and pinworm. It has been used to treat dysmenorrhoea and malaria. Some of the biochemicals have been used to treat the autonomic nervous system disorders like epilepsy. The leaves have been used to promote wound healing and to decrease the pain of neuralgia and rheumatism. Tinctures have been taken in whisky to treat asthma.

The oral history of most of the aboriginal medicine has been lost. Just one break in the chain of memory of one generation will do this. This loss affects the whole of mankind in its present time and for the future.

Panacea

The medicine woman of the First Nations brought me to see one of their eight major medical pharmacopoeia growing in an ancient natural habitat. We walked a long way, one following the other. We stepped over a sleeping snake. We sank into a sphagnum bog. Then we came to a partially shaded embankment, an old moraine from the last ice age.

She quickly and elegantly plucked a root of goldenseal, *Hydrastis canadensis*. She skinned it with one quick pull of her nail. She gave me the golden thread to suck and hold in my mouth. "Now," she breathed. "That, Panacea! You give me enough that one"—she pointed to the bank—"and I will cure AIDS." The root had a bitter taste on my tongue, and its memory remains. She is their last medicine woman. When she dies all that knowledge, too, will be lost.

ECOFUNCTION

The wafer ash, P. *trifoliata*, has an extremely clever antifeeding strategy. Being a small tree, it could easily be on the endangered list due to browsing by large mammals. The glandular cells that are found all over the plant hold their terpene oils under tension so that each gland can become explosive like the citrus glands in an orange skin when peeled. If a twig, leaf, or branch is crushed, a mixture of terpene, orange-smelling α-limonene, and other volatile oils are released into the air. These are antifeeding compounds to all mammals that are fur bearing and also to large birds.

These mammals also get a fine misting of an aerosol on their coat that is in fact a wetting agent. It acts as a skin irritant. Then another chemical called xanthotoxin follows up. This toxin produces phototoxicity in the animal under bright light when the xanthotoxin is groomed from the fur. These are lessons in avoidance therapy that are listened to by the animal kingdom with unswerving obedience.

The wafer ash, *P. trifoliata,* walks a fine line. On one hand the tree tries to protect its life with antifeeding chemicals, and on the other hand it needs to attract insects for the pollination of its flowers. The smaller flying insects and honeybees are attracted to the flowers by one of the strongest lactone fragrances in the forest. The insects follow the truly magnificent fragrance to collect nectar from the blooms in a flow that lasts a little longer than a week in cooler weather and less than a week in bright, hot, sunny conditions. These insects and, in particular, honeybees, also collect another chemical bounty from the flower. It is a strange-looking tropate oxide that is also found in the *Datura* of the Solanaceae family. It is called scopolamine *N*-oxide. It helps the insects fine-tune their motions in flight and gives them a feeling of well-being. These medical nectars are evaporated into a honey in the hive by the worker bees. This honey is always compartmentalized in comb sections that are kept separate from other honey collected at the same time during the season.

The wafer ash, *P. trifoliata,* is the host species to some of the more dramatic butterflies in the north temperate garden. These large butterflies, with their bright markings, depend on the toxic arsenal of chemicals in the leaves of the wafer ash for protection against avian predation. The drifting eastern black swallowtail, *Papilio polyxenes;* the hilltop flyer called the anise swallowtail, *P. zelicaon;* the orange dog also known as the giant swallowtail, *Heraclides cresphontes;* use it, as well as the ruby-spotted swallowtail, *Priamedes anchisiades,* which is a subtropical butterfly. The little sickle-winged skipper, *Achlyodes thraso,* uses the *Ptelea* leaf to hang its grass green chrysalis from a silken thread in open hilly country.

Wafer ash, *P. trifoliata,* uses a group of compounds called phytoalexins in a defense mechanism in response to attacks by

North American native *Ratibida columnifera,* prairie coneflower. Its summer-long flowering makes it an ideal companion to *Ptelea trifoliata,* wafer ash.

insects or fungi. These compounds maintain the wafer ash in a disease-free state throughout its life.

BIOPLAN

All of the different species of *Ptelea* and their sports and cultivars should be maintained in arboreta in Mexico, the United States, and Canada. These species should also be reintroduced into their wild habitats in their full range on the face of North America. Then these species will be able to help maintain the butterfly populations and to stabilize their migrations and range. Butterflies are pollinators of the native herbaceous plants and trees. To protect these species is to ensure the survival of many of our native species of plants.

Since it is becoming less and less acceptable to poison our environment, the citrus growers in the southern parts of the country could plant hedgerows of the wafer ash, *P. trifoliata,* as attractant hosts for the beautiful giant swallowtail, *Heraclides*

Opuntia macrorhiza, Missouri prickly pear cactus. It will endure as much drought as will *Ptelea trifoliata,* wafer ash, for both thrive in dry conditions.

cresphontes. There is presently much spraying to kill this butterfly in citrus groves. Alternative feeding for this butterfly species using *P. trifoliata* in a similar manner to the use of mulberries, *Morus,* species in cherry orchards would be a much less toxic alternative. A higher price could be obtained for organic citrus produce.

DESIGN

Any fragrance garden should have the wafer ash, *P. trifoliata,* or either of the more rare *P. nitens* or *P. polyadenia,* which are both very fragrant. These species can be placed in positions where the leaves can be brushed in passing to release a degree of orange fragrance into the airways around the home in addition to the heady honeysuckle fragrance of the flowers.

In warmer gardens of zones 5–9 and possibly in zone 4 the *P. baldwinii* (syn. *P. augustifolia*) with smaller narrower leaves and larger individual flowers can be used. All of these species, too, could be part of a butterfly border.

Gravel gardens are coming into fashion in Europe and now in North America. The wafer ash, *P. trifoliata,* is an ideal specimen tree for such a garden. They can also be used to great advantage as a smaller tree in a large rock garden.

There is also a very fine, golden-leafed cultivar of *P. trifoliata* called *P. t.* 'Aurea'. The golden palmato form contrasts extremely well with the purple foliage of the smoke tree, *Cotinus coggygria* 'Royal Purple'. There is also a fine contrast with the blue-green foliage and pink flowering cascades of *Tamarix ramosissima* 'Pink Cascade'. The native *Clethra acuminata,* with its polished, cinnamon brown bark, can be used for a more delicate visual contrast.

For the tiny garden, or the limited space of irregular suburban gardens, the erect form of *P. trifoliata, P. t.* 'Fastigiata', is ideal. It matures to a columnar 24 feet (7 m) with noninvasive roots adding dignity and vertical lines to any house and garden.

Block plantings of the heartleaf crambe, *Crambe cordifolia,* which is fragrantly in flower all of June into July, can be used with *P. trifoliata* in richer soil. Groupings of the ground clematis, *Clematis recta,* with its sweet, nostalgic, fragrant white flowers, can be used in poorer dry soils. For the more exotic touch, in the warmer zones of 8–10, the African iris, *Dietes vegeta* 'Alba', can be used around the base of the tree.

Throughout the late fall and into the winter months, the swags of seeds on *P. trifoliata* are always attractive and are seldom passed without comment.

Ptelea Species and Cultivars of Merit

SCIENTIFIC NAME	COMMON NAME	ZONES
Ptelea baldwinii	Baldwin ash	5–9
P. b. var. *crenulata*	Crenulated wafer ash	4–9
P. nitens	A rare aromatic tree found in southwestern U.S. in 1912	5–9
P. polyadenia	A rare aromatic tree found in southwestern U.S. in 1916	5–9
P. trifoliata	Wafer ash	3–9
P. t. 'Aurea'	Golden wafer ash	3–9
P. t. 'Fastigiata'	Fastigiate wafer ash	3–9
P. t. 'Pallida'	Pale wafer ash	3–9

Ptelea trifoliata, wafer ash. It can take a place in a North American medicine walk.

Quercus alba, white oak, showing a massive dominance in the northeastern corner of its American forest

Quercus

OAK

Fagaceae

Zones 3–10

THE GLOBAL GARDEN

Like the dog, the oak has been man's companion, it seems forever. Sanskrit has the oldest recorded use of the word for oak. It is *daru,* just a hair away in time from the word *dair,* which is the second letter of the Ogham script, also meaning oak. Ogham was an early Irish sophisticated written language based on trees. This script is found throughout Ireland and Europe. It is also found in North America and perhaps explains the reason why the oak has been so treasured on this continent by the aboriginal peoples.

In ancient Europe and stretching eastwards, the oak has had deep religious and magical meaning. This tree was so honored that, if a man harmed an oak, his navel was nailed to the bark and he was forced to walk round and around the tree, unraveling his intestines in a death walk. Later, oak became a source of building materials and communal commonage for mast. This fed animals and people, too, in a pinch. In the Americas the oak has always been an important winter food obtained from the sweeter acorns, which grow in abundance and in great variety. The important medicines of the oak were prepared ceremonially by the medicine men of the aboriginal peoples that began with *Na'bunong,* meaning eastward. Perhaps their mythical past was in the east.

The Druids, members of the ancient Celtic priesthood, drove their power base on science, mathematics, and magic. The priesthood was inherited from father to son in a society that still touches us into this millennium. The chosen names for their sacred oak groves, like "Kildare," are the coinage of today. The height of druidic magic was the control of lightning in their place of worship. The oak, with its large, open, internal pores, is hit by lightning more often than any other tree. The Druids capitalized on this powerful fact. The oak was also dubbed by the high priests to be magical because it harbors a parasite called mistletoe, *Viscum album,* in its branches. Mistletoe, with oak as its host, carried the most powerful and magical medicines of the ancient world. Indeed it appears to have lectins, which are now so hotly debated by scientists. Lectins specifically target cancer cells and kill them. So the act of the druidic priesthood of a kiss under the mistletoe has more meaning than simple affection.

The Druids have had the last laugh. They believed that the oak had the ability to listen and to hear. Pre-Celtic mythology is littered with deaf kings, listening ancient oaks, and singing harps. Cutting-edge research has recently demonstrated that trees can hear. Their listening range is 2 kilohertz and about 70 to 80 decibels, which is a little bit louder than the normal human voice. The listening response in plants is the synthesis of gibberellic acid. This, by the way, is an acid very similar in chemistry to the anabolic steroids responsible for muscle growth and sealing the fate of many an Olympian when taken illegally.

Quercus is a very large genus of some six hundred species that keeps growing. The opportunistic oak, as conditions change, will hybridize quite happily. Oaks are deciduous and sometimes evergreen. They range from being truly massive in size to being almost like a shrub. Oaks are found widely distributed in the northern hemisphere of the global garden as well as the tropics. Some oaks like the Chinquapin, *Quercus muehlenbergii,* grow in the cloud forests of Mexico between Monterrey and Ciudad Victoria. They track up through central Texas, swarm the Great Lakes, and end up in Ontario, Canada.

A special, visual landscape exists in North America. It is a relic of aboriginal horticultural practices. The landscape is called

Stylophorum diphyllum, wood poppy, endangered and beautiful. It is a member of the poppy family, Papaveraceae, and is found in the rich woods of Canada through Pennsylvania to Missouri.

Of the many fungi associated with *Quercus*, oak, the *Boletus* species of mushrooms are some of the most commonly seen in North America today

savannah. It is composed of prairie grassland that is covered by 10 to 20 percent oak. It is *Quercus macrocarpa,* the bur oak, that populates the U.S. Midwest. The Canadian savannah tree is *Q. velutina,* black oak, or *Q. alba,* white oak. The more rare *Q. bicolor,* the swamp white oak, and also *Q. macrocarpa,* bur oak, savannahs are still found in Canada. The aboriginal peoples maintained these open savannahs as a source of game. They controlled growth on a massive scale by April and November flash fires. Recent paleoecological research on soil strata attest to savannah horticultural practices all over eastern North America.

Another strange DNA fingerprint of oak has been left behind by pioneer practices. This is found in North America as well as in Russia. The logging of *Q. macrocarpa,* the bur oak, has only been for straight trees with long trunks for the timber trade. Consequently, what was left behind are the crooked, misshapen bur oak species that have propagated themselves selecting for warp. These are known as retrograde oak species, the crooked being common with the straight trunks becoming exceedingly rare in the global garden.

All of the oaks, the *Quercus* species, are sunbathers. They soak up the rays like no other tree. The more sun, the better they like it. A savannah allows them to do this. Sun spells success for the oaks because they set about their business of trapping electrons by photosynthesis into huge crops of protein-laden acorns. Some species take two summers worth of sun to manufacture their acorn crop. Others take more time. Some years, with more intense sunshine, large crops of acorns are produced. The aboriginal peoples were aware of these yearly rhythms and stored sacks of acorns in the natural refrigeration of mud under cool, running water for years of scarcity.

Despite a great variation in leaf shape, all oaks produce acorns. Each acorn is like a cup and saucer. The shape and size of the cup with the decoration or "filigree" and depth of the saucer are the most useful items for oak classification. In recent years three new oak species have been found in Canada. These are *Q. shumardii,* the Shumard oak; *Q. ilicifolia,* the bear oak; and the *Q. ellipsoidalis,* Hill's oak.

In contrast to the increases in Canada, the two native English oaks, *Q. petraea,* the Durmast oak, and *Q. robur,* the English or common oak, are in decline, especially the latter. An immigrant American fungus called *Microsphaera alphitoides,* an oak mildew, is having a heyday spinning mycelial growth on leaves and acorns through the British Isles with no end in sight. The excessive moisture of global warming and high pollution levels are not helping. However, there is an old saying, an oak must have 300 years to grow, 300 years to live, and 300 years to die. A lot can happen in a thousand years, especially for an oak.

ORGANIC CARE

The oaks can be divided into the white oak group, the red or black oak group, and the evergreen species of oak in the tropics. These latter ones are the huge California live oaks, *Q. agrifolia,* seen on visits to that state, and the Virginia live oaks, *Q. virginiana,* of Cuba and the Deep South of the United States.

The white oaks like a neutral soil pH 6.5–7.5. This includes the white oak, *Q. alba;* the swamp white oak, *Q. bicolor;* the Shumard oak, *Q. shumardii;* and the English oak, *Q. robur.* These trees prefer a rich, porous soil overlying a dolomitic subsoil into which they can sink their considerable roots and remain drought resistant. These oaks, like all fruit producers, need calcium for sweeter nuts.

The black oak group prefers a deep, rich soil somewhat acidic with a pH 6.0–6.5. This includes the black oak, *Q. velutina;* the pin oak, *Q. palustris;* the chestnut oak, *Q. prinus;* and the popular scarlet oak, *Q. coccinea.* The heat-loving southern oaks such as the blackjack oak, *Q. marilandica,* prefer more acidic soils, again from pH 4–5 and lower.

From the acorn all oaks will grow exceedingly well. The acorn has internal embryonic leaves or cotyledons that are ready to expand and begin growth as soon as the acorn touches the ground. For this reason, acorns that are destined to be hand-planted should receive some care. They can be put into a pail of water. The weevil-infested floaters can be discarded. The remaining acorns should be kept in humid conditions and should not be overheated. They will turn a tan color a few days after harvest. All the acorns of the white oak group are, then, ready for planting outside immediately.

The black oaks must have their acorns stratified to break dormancy. They can be placed in damp peat moss and sand or be potted and kept for up to 220 days at 36°F (3°C). The acorns can be planted outside in nursery beds after stratification. They should be planted on their bellies for ease of germination. They may have a light ¼-inch (.6 cm) covering of soil. This can be covered by 2 inches (5 cm) of dried leaves to prevent soil desiccation. The first year, seedlings will rapidly grow to about 5 inches (13 cm).

Oaks are difficult to transplant. They resent being moved and like to have their considerable network of adventitious roots to be kept intact. The best way to do this is transplanting a one- or two-year-old tree to its final site. In the colder zones 2–5, this should be done as soon as the ice has gone as close to the month of April as possible. This is when the water in the soil begins to recede due to the decrease of hydrostatic pressure. The young oak roots can follow the water as it moves deeply back into the subsoil. For the warmer zones 6–10, a fall transplantation of the oak is usually quite successful, especially when the young tree is fully dormant and the soil is tamped well around the planting hole so that a catch basin for rainwater is formed.

A *Viburnum prunifolium,* black haw, variant that borders oak woods and acts as a feeding, resting, and staging ground for the songbirds of North America

The soil for the planting hole for any of the oaks can be enriched with either bonemeal that has been steamed or old soup bones placed deep into the hole. These act as a slow-release phosphate fertilizer for the young tree. A surface mulch of 4–6 inches (10–15 cm) of aged horse manure around the catch basin helps vigor in the leaves and keeps the roots cool while the tree becomes established in its open, sunny position.

The growing oak is a pastoral metropolis producing pollen, nectar, honeydew, and its crop of acorns. Consequently the oak

Quercus macrocarpa, bur oak, taking a seedling's view of its nursery ground in a mature white cedar, *Thuja occidentalis,* woods

harbors literally hundreds of kinds of galls incorporated into the leaves, twigs, and flowers. These galls are homes to many species of flies, nonstinging insects, oak lace bugs, and leaf-eating insects, making the oak a happy hunting ground for predation and prey.

MEDICINE

The medicine of the oak is to be found underneath the bark. For the most part it is a chemical called quercitrin that supports the tree's ability to trap and use sunlight in the shorter wavelengths. This and a similar biochemical, quercetin, are important vasoactive drugs that help control blood pressure in man. They increase blood pressure by reducing the diameter of the arteries, which, in turn, speeds up blood flow. These are the same drugs found in the Egyptian lotus, *Nymphaea caerulea.* These drugs were used for fertility rites throughout ancient Egypt. This was the "Viagra" of the pharaohs.

The Cayuga peoples used American elm, *Ulmus americana,* and swamp white oak, *Quercus bicolor,* for the treatment of broken bones. A 2-inch-long (5 cm) strip was taken from trunk callus tissue of each tree. Callus tissue occurs on these trees when wounded. The callus should be hemispherical, clean looking, and about to close in on itself on the trunk. These strips are boiled in 6 quarts (5.7 l) well water. The decoction is drunk as often as needed.

Much use was made of the common polypody fern, *Polypodium virginianum,* by the Seneca peoples. They commonly treated cholera with a concoction of swamp white oak, *Q. bicolor;* the polypody fern, *P. virginianum;* and the fresh green needles of northern hemlock, *Tsuga canadensis.*

The Mohawk peoples used a decoction of the bark of white oak, *Q. alba,* in the treatment of horse distemper.

The Iroquois used red oak, *Q. rubra,* in an interesting treatment for the curing of ruptured navels. Callus bark in its rounded, healing growth, was scraped off the tree. This was dried and powdered into a fine dust. This was put on the navel.

Ohaguro

One of my first memories of childhood in Ireland was that of being introduced to a family friend who had recently been married. She smiled down at me with a great big toothy smile. Her new Irish husband stood proudly by her side. I stared in disbelief at her teeth, they were all black. She, in her Japanese heart, was showing a mark of old-fashioned fidelity to her new husband.

Ohaguro is a two-thousand-year-old practice in Japan. In Central and South America, in parts of Africa, and among some peoples of India, this custom exists also. Tannin from oak is mixed with iron to make ferric tannate, which colors and protects the teeth . . . it is only another ancient way of saying, "I respect you. I love you."

The Appalachian women used the bark of white oak, *Q. alba,* as a relief for tired and swollen feet. Strips of bark were boiled in well water and were cooled to lukewarm. The bark was removed and the warm water was used as a foot bath.

The Chippewa peoples used the bur oak, *Q. macrocarpa,* for lung trouble and pains in the chest. They also used the bur oak for the treatment of the circulatory system in a special ceremony using very precise measurements of medicinal liquid extracts poured into a birch bark volumetric measure. From this, precisely one swallow of the drug was taken. The extract was made from the inner bark of bur oak, *Q. macrocarpa,* and the red oak, *Q. rubra.* Both of these were dried and powdered. The inner bark of trembling aspen, *Populus tremuloides,* together with a mixture of equal amounts of root, bud and blossom of balsam poplar, *Populus balsamifera,* were rolled in the palm of the hand, to bruise them lightly. They were then mixed with the root of Seneca snakeroot, *Polygala senega.* Into 1 pint (.5 l) of well water the four tree mixtures were placed at the four points of the compass: a pinch of bur oak to the east, a pinch of red oak to the west, weeping aspen to the north, and balsam poplar to the south. Then on top of each pinch, a further pinch of the fabled, powdered Seneca snakeroot was placed, repeating the words of direction. This was allowed to steep. This medicine was extraordinarily powerful. It contains vasoactive drugs, corticosteroids, and antibiotics. One swallow was taken. This dose was repeated again in exactly one hour. It was like a chemical shock treatment.

The oaks are very high in tannins. This is also the basis of them being used as a medicine. Members of the white oak group contain about 10 percent tannin; the red oaks contain 20 percent tannin; and the evergreen oaks, also called the cyclophora group, have 10 to 20 percent tannins.

The Seneca peoples used a cranberry bush bark decoction in water as a strong laxative. They used the bur oak, *Q. macrocarpa,* as a countermedical antidote to diarrhea. The Seneca peoples also used the leaves of slippery elm or red elm, *Ulmus rubra,* and swamp white oak, *Q. bicolor,* in the treatment of catarrh. Mature leaves of both species were collected, strung, and dried. They were then rolled together into a cigarette. This was smoked, making the smoke run through the nostrils. The smoke was trapped by the mucosa of the sinuses. A clearing cough was ensured.

The oak, particularly the black oak, *Q. velutina,* was used by the aboriginal peoples of North America as a toothpaste. The powdered bark of black oak was mixed with the bark of black alder, *Alnus glutinosa.* To this was added bayberry, *Myrica pensylvanica,* and wild ginger, *Asarum canadense.* A paste was made by adding equal proportions and rubbing it on to the teeth with a stick brush or by finger. This mixture also prevented tooth decay.

Monarda didyma 'Alba', the rare white or *alba* form of Oswego tea found in virgin forests. It helped, through the Boston Tea Party of 1752, to redefine the American view of public taxation.

A mighty *Quercus macrocarpa,* bur oak, that even a modern farmer could not put to the sword

ECOFUNCTION

Throughout the world and especially in North America, the oak has fed an extraordinary array of creatures, from insects and butterflies to birds such as blue jays and wild turkeys and domestic fowl. The acorns were also mast for pigs, goats, opossums, foxes, raccoons, and bears. The sweeter acorns were used as a vegetable by the aboriginal peoples.

The more bitter acorns were boiled twice either in water or a wood-ash, lye producing, water bath that released the tannins from the acorns. They were then ground into a flour. A hard, dry bread was baked from this flour that has the taste of a cross between cornmeal and peanut butter. An oil, similar to olive oil, was also pressed from the acorns of some oak species. Pioneer farmers marked their pigs with their own individual marking system. These pigs were let loose into oak forests where they fed on the acorn mast in the fall and early winter. The meat from these pigs had a fine flavor. The meat and bacon products from oak mast–fed pigs was said to keep for a long time and did not have a tendency to go rancid.

All species of oak carry a strong attractant for squirrels. And since the squirrel is nature's forester, it is in the interest of the oak's survival as a species to have these rodents nearby. Acorns have a special sugar called acorn sugar or *d-quercitol.* It is a form of inositol sugar that is also very similar to glucose. Squirrels crave this sugar; they eat some acorns and bury others in hordes for a rainy day. They have very poor memories and that is how an oak forest is born. An oak forest generated by squirrels is one of the highest genetic quality because the squirrel rejects any damaged acorn. So the forest is grown from only the best.

The oak has another card to play when it comes to protecting itself against the feeding practices of larger animals. Tannins are found in the mature leaves. Repeated feeding on oak leaves, especially at times of stress for the oak, like drought, is toxic. The tannins go into the stomach, across the intestinal wall, and directly into the blood. This can be fatal. Browsing deer in the springtime take a little nibble on buds and on the very young leaves, which are still low in tannins. They will not return to eat the mature leaves from midsummer into the fall.

The feeding pattern of the Kermes oak, *Q. coccifera,* from Portugal and grown in England since the early 1600s has produced a strange industry somewhat like the Chinese silk growers. This oak plays host to the Kermes insect, from whose body one of the most extraordinary dyes in nature is obtained. It is called cochineal. A water extract from the bodies is a deep fluorescent purple-red that changes into a pure scarlet on wool, linen, cotton, and silk fibers. Approximately seventy thousand insect cadavers make up around a pound in weight. This number of insects is capable of dyeing and redyeing a huge amount of cloth. The grooming of this 6-foot (2 m) tree is as easy as it is lucrative for the dyers' trade.

Pulsing through the vascular system of many oaks is a brilliant yellow dye that acts like a suntan lotion. This dye absorbs light in the high-energy, invisible ultraviolet end of the light spectrum in the 350 and 258 nanometer range. On one level, because the oak demands maximum solar exposure for good growth, this dye, like sunscreen, acts as an overall protector of the tree. But on another level, the dye is composed of two biochemicals, quercitrin and quercetin, both of which are capable of being energized to a higher energy state by photons, especially ultraviolet photons. It is possible that the dye composed of these two biochemicals is functioning as a liquid phase electron storage for the oak for photosynthesis. If this is the case, the oak gives a new meaning to molecular solar receptors and collectors outside of the chloroplast.

There are many species of mushrooms associated with the oak. Some of these are *Boletus, Lycoperdon, Morchella,* and *Russula* ssp. In addition, a great number of butterflies use the oak as a feeding host across the North American continent. These are various true skipper butterflies, hairstreaks and brush-footed butterflies like the California Sisters.

Some oaks produce acorns that can be eaten fresh like a nut. Generally these are oaks growing in warmer areas, the exception being some occasional sports of the white oak, *Q. alba,* and the bur oak, *Q. macrocarpa.* The warmer oaks are the dwarf rock chestnut oak, *Q. prinus pumila,* and the sweet acorn oak, *Q. ballota,* of Morocco, which is also a source of oil in that country.

BIOPLAN

The oak forests that are being cut across the planet are not being replaced. These species are going into decline as quality forest species. This is especially true for the west coast of North America, where the blue oak, *Q. douglasii,* and the poorly treated California white oak, *Q. lobata,* together with the Garry oak, *Q. garryana,* which extends into the west coast of British Columbia, are being extirpated without thought. Quality oaks are not being identified and preserved, nor are their acorns salvaged for propagation.

The oak is the fortress of the north temperate forest. These species, together with their deciduous companions, produce a

Sisyrinchium angustifolium, pointed blue-eyed grass, an alkaline footprint of the ancient aboriginal bur oak, *Quercus macrocarpa,* savannahs that graced the face of the North American continent once upon a time

global cycle of atmospheric gasses as they switch on growth in the spring and switch it off in the fall. The gasses are carbon dioxide and oxygen together with other volatile chemicals. Taking a holistic view, the annual spring trough of carbon dioxide with its winter peak of dormancy is a respiration pattern of the planet itself and is common to all living things. The off position usually means death. The respiration pattern of the planet is beginning to plateau, the peaks and troughs are flattening out. The forests must come back. And from an acorn an oak forest will grow.

Oak forests can be replanted as epicenter forests. These are forests that are begun with four or five top-quality genetic stock trees. These trees are pampered for their first 10 years, protected from grazing and girdling, after which time they, by growth and natural selection, begin a forest. Given a little nitrogen encouragement and some protection, oak species grow remarkably rapidly. Within a few years they will hold pace with the growth of conifers nearby.

The bur oak, *Q. macrocarpa,* can be planted into unused farmlands to form savannahs. This oak species is fire resistant

Blephilia hirsuta, wood mint. This fragrant mint is found in the dappled shade of virgin forests as far south as eastern Texas. Its beauty is unparalleled when massed.

after about 15 years of growth. It could also be planted within reforested areas as a fire shield or as a fire boundary.

In regions where vineyards are being developed, small areas of farmland should be bioplanned for swamp white oak, *Q. bicolor,* and white oak, *Q. alba,* for barrels for wine storage and aging. Presently there is a shortage of oak in the cooperage trade.

Botanically speaking, an acorn is a true nut. As such, this food is high in the B complex of vitamins and high in protein. This food only needs a little commercial imagination, like the raw olive in pickle brine, to leach out the bitterness and food industrialists could take the acorn to the same heights as the soya bean. The oak could then come the full circle, back to being a clean food source for mankind.

The following selection of oaks are ideal for urban reforestation on the North American continent: the bur oak, *Q. macrocarpa;* the black oak, *Q. velutina;* the white oak, *Q. alba;* the scarlet oak, *Q. coccinea;* the red oak, *Q. rubra;* the English oak, *Q. robur;* the Hungarian oak, *Q. frainetto;* the Mongolian oak, *Q. mongolica;* the pin oak, *Q. palustris;* and the swamp white oak, *Q. bicolor.* There is also the durmast oak, *Q. petraea,* which is excellent for maritime areas. These oaks grow well in zones 3–9 except for the scarlet oak, *Q. coccinea,* which is a little less hardy and will grow in zones 4–9. The giant Shumard oak, *Q. shumardii,* is ideal for city parks or street allées.

DESIGN

Vegetable gardeners may look to the white oak, *Q. alba,* as an indicator tree. When the leaves are ready to unfurl, it is time to plant the corn.

For a large garden or an estate, a well-placed oak of any species is a tree of beauty. However, the majority of gardens are small and urban. Happily, there are a good number of oaks that can be considered as ideal for the suburban garden. There is the native bear oak, *Q. ilicifolia,* for zones 4–9. This is a small tree that will grow in poor soil and has a wonderful pink flush of growth in the spring. There is also the shin oak, *Q. mohriana,* zones 5–9, and the Georgia oak, *Q. georgiana,* zones 5–9. Getting more easy to find are a number of popular cultivars of the English oaks, *Q. robur,* such as *Q. r.* 'Pendula' and *Q. r.* 'Fastigiata'. The latter has a handsome purple-leafed cultivar, *Q. r.* 'Fastigiata purpurea', which can be tucked into any small corner. It would also combine well with groupings of the native wood poppy, *Stylophorum diphyllum,* a brilliant yellow flower with sculptured foliage. Any small oak could also be the trellis for the native vine called the hog peanut, *Amphicarpa bracteata.*

A source of water, as a birdbath, should be near the oak. There may be a piece of statuary, an irregular stone, or interesting wooden stump, and then you have produced a little urban forest haven. There is also considerable opportunity for the urban gardener and entrepreneurial horticulturist to develop more native North American species for the ornamental trade.

Quercus Species and Cultivars of Merit

SCIENTIFIC NAME	COMMON NAME	ZONES
Quercus alba	White oak	3–9
Q. bicolor	Swamp white oak	3–9
Q. coccinea	Scarlet oak	4–9
Q. frainetto	Hungarian oak	5–9
Q. garryana	Garry or Oregon oak	5–9
Q. georgiana	Georgia oak	5–9
Q. macrocarpa	Bur oak	3–9
Q. mohriana	Shin oak	4–9
Q. mongolica	Mongolian oak	3–9
Q. palustris	Pin oak	4–9
Q. petraea	Durmast oak	5–9
Q. robur	English oak	4–9
Q. r. 'Pendula'	Weeping English oak	5–9
Q. r. 'Fastigiata'	Fastigiate English oak	4–9
Q. r. 'Fastigiata Purpurea'	Purple fastigiate English oak	4–9
Q. shumardii	Shumard oak	4–9

Trillium cernuum, nodding trillium, bows to the forest above.

Sambucus, elderberry. Small shallow ponds, rivers and streams, and even the odd roadside ditch are the places to find the elderberry tribe. Look in September because the fruit, either red or black, will position the species in its habitat, which should be less damp at that time.

Sambucus

ELDERBERRY

Caprifoliaceae Zones 3–10

THE GLOBAL GARDEN

The story of the elderberry has never been told. In the past it was part of the tapestry of living memory, fingertip knowledge shared from grandmother to child. Like the wild strawberry, it has dropped into oblivion, uses forgotten, so much so that in this age of information many a gardener is found looking for elderberries in places they would never exist. A sad fact of life is that the elderberry is cast aside, to be poisoned again and again as a so-called silvicultural weed by agricultural authorities.

The peoples of the ancient world had pet names for the elderberry, calling it *akle, rixux,* or *ixus.* The Romans called it *ebulus* after *ebullire,* meaning "to bubble out." The Celts named it *crann troim,* the laden tree. The Egyptians, Africans, Chinese, and Japanese named and treasured it. It is part of the medical pharmacopoeia of the First Nations of North America. It is still in use in Mexico, bought and sold in their local herbal trade.

There are about twenty or so species of elder or *Sambucus* in the global garden. They stretch in growth from the semitropics up into the colder zone 3 in the north temperate regions of the world. Usually the tropical species are small to medium-sized trees. They tend to get smaller in stature as they head north. With this comes a change of habit; suckering becomes more common in the northerly species. In the coldest habitats the genetic material of *Sambucus* seems to twist and turn in an agony of survival. Out of this come the finest cultivars and the strangest of sports for the gardening community to adopt and enjoy.

Trailing around northern Africa into the Mediterranean and occasionally popping up in England is the one poisonous *Sambucus* perennial species called Danewort, *Sambucus ebulus.* Apart from this, they can be classified into two groups, the species that bear a crop of red berries and those that have blue-black fruit. The latter seem to be endowed with magic and medicine all over the global garden. The former, those with the fall crops of red berries, are a closely inspected and coveted bounty for wildlife.

Belief in elder magic was so strong in many northern European countries that the cuttings from these species were never burned. The local people strongly believed that the souls of the dead were still harbored in the branches. Consequently, they would be lost along with the cuttings if burned. These same people are still seen to pay homage to the elder with a quick doff of the hat and a bow of the head in respect, as a formal greeting.

The *Sambucus australis,* the Australian elderberry, is found in the southern portion of South America. *S. valerandi* is found in Asia, South America, and southwestern Australia. *S. javanica* is found in Taiwan and Malaya. The evergreen, *S. canadensis* var. *laciniata,* grows in Florida together with the tree forms of *S. caerulea* (syn. *S. glauca*) and the two semitropical races, *S. c.* var. *neomexicana,* and *S. c.* var. *velutina* (syn. *S. californica*). Throughout the North American continent there is *S. canadensis,* the American elder, especially found hugging the eastern maritime coastline. The English *S. nigra,* with its cultivars, has also found a home in Canada, as has its close relative the red-berried *S. racemosa,* the European red elderberry. The handsome American red, *S. pubens,* is seen occasionally with its hoard of scarlet berries, but seems to be getting quite scarce in recent years. The eastern, *S. pubens,* has a string of red representatives on the west coast. These species are more geographical in variation. The more northern on the Pacific Coast are *S. callicarpa.* Getting closer to the Rocky Mountains and into California, the

S. melanocarpa is found. This gives way to *S. microbotrys* of the California area.

ORGANIC CARE

More often than not the elderberry is propagated by both stem and root cuttings, but the elderberry can also be propagated by seed.

Softwood cuttings of all species elderberries can be taken in April, May, or June. A number of 6-inch (15 cm) tips are hand-snapped at a convenient node. One pair of leaves is left on the cutting, which is potted up and placed in a shaded area. The cutting is well watered and kept damp for 24 to 48 hours. After the cuttings become turgid, they can be slowly moved out into the sunshine. The soil is kept damp for a further three weeks, at which time rooting will have commenced. The cutting is not allowed to dry out for a further two months. It can then be planted out in the nursery bed. This will be some time in midsummer. The cutting is allowed to grow on.

Many of the elders produce an L-shaped root sucker. The stem can be cut back to 6–8 inches (15–20 cm) in length, retaining the root piece, which can be trimmed, too, if it is too long. This is potted up and kept well watered. It will grow on in a month or so. Root suckers will develop rapidly into plants if taken in spring, but they can be cut and propagated at any time in the growing season, taking a little longer to mature. The suckers should also be well watered during their first season.

All species of *Sambucus,* elder, can be propagated by seed. The fleshy seed is technically called a drupe. This ripe fruit can be squeezed to produce three to five tiny seeds or nutlets. All of the seeds are remarkably similar for all of the elders, and all have a rough, hard, outer seed coat. This coat must be softened to allow the embryo to commence growth. The elder embryo seems to have a somewhat variable dormancy time that corresponds with the length of the local season of growth.

When the fruit begins to drop in the late summer or early fall, the seeds are ripe for collection. The heads of fruit are cut and collected in a flat basket. Later on, the fruits are easily stripped from the heads by hand. They are collected and spread out in a single layer to prevent heating and fermentation.

In a day or so, while the fruit is still soft, it can be run through a sieve, or an equal volume of water added to the fruit. It can then be put into a blender and blended for the minimum of time. The seeds are washed.

The damaged or sterile seeds will float away. The seeds can be strained in a filter paper funnel. A coffee filter works nicely. The seeds are then spread out on toweling or screen and allowed to dry. The dried seeds can be stored at 41°F (5°C) in a refrigerator in a glass jar for up to two years with little or no loss of viability.

Nature needs to be mimicked to break dormancy. Normally, birds perform the task of seed gathering. They eat the berries. These pass through the stomach with its strong acid contents, which scores the hard seed coat or endocarp. The seeds continue on their passage through the intestines and are purged from the bird in readiness to germinate the following spring.

To break dormancy by mimicking nature, the seeds first need to be warm stratified for 60 days at 70°F (21°C). The seeds are then subjected very carefully to a five-minute soak in concentrated sulfuric acid. The seeds are rinsed. They are then subjected to 90 days at 41°F (5°C). This time can be extended. The seeds are then planted out into a nursery row at a rate of two seeds per 1 inch (2.5 cm). The seeds are lightly covered with ¼ inch (.6 cm) of soil. This in turn is covered with one layer of Reemay for shading as the seeds germinate. At all times the nursery row is kept damp.

In the warmer zones 6–10, where the dormancy aspect of seed growth is not quite so acute, the seeds can be collected as they ripen. They are extracted, put through the five-minute concentrated sulfuric acid bath, washed and planted out in a nursery row. The seeds may decide to take two full growing seasons to germinate, so the nursery row should be well marked.

The seeds grow by means of epigeal germination, which is

slow. The process takes six weeks to produce a small seedling with its first true leaves. By the end of the summer the plant looks like a miniature tree, with smaller buds, leaves, nodes, and internodes, all of which are tinged with purple. After three years the plant metamorphizes itself into an adult elder. It is then ready for transplantation into its final location. This process of collecting seeds to transplanting the seedling trees takes five to six years.

Elders will grow in a wide range of soils and pH. They grow extremely well in acidic soils and will also tolerate alkaline conditions. Elders like to have damp, cool roots that can be supplied by a soil mulch. They also like to have sunny heads for fruiting. If the elder are to be grown in the kitchen garden for fruiting, the soil should be amended with peat moss or well-aged manure that is placed deeply in the planting hole. One pint (.5 l) of steamed bonemeal and the same of woodash will pay dividends in fruit. The trees can also be mulched with wood chips, grass clippings, or spoiled hay if the site is a little dry. The mulch should be maintained 3–4 inches (7.5–10 cm) thick year round.

The elderberries grown for fruit, wine, or flavored drinks, *S. canadensis,* the American elderberry; *S. nigra,* the common elderberry; and *S. glauca,* the blue elderberry; all respond to a rotational pruning once every two to three years, especially in the more northerly locations. This practice considerably increases flowering and fruiting. When some cultivars of *S. canadensis* get too straggly they are spring pruned to ground level to within three to four buds from the ground as a regeneration technique.

For many reasons the elder is remarkably free of pests. The entire clan seems to be immune to infection and attack throughout the growing season. As a kitchen fruit, they are the most carefree and easily grown fruit for the gardener. There is an occasional berry worm in the odd panicle that is easily seen by its white web. The true enemy of the elder is severe drought; however, they seem to spring back to life if there is rain the following year.

Sambucus canadensis, American elderberry, with an umbel of ripe fruit. This is the more common elderberry of eastern North America.

Chelone obliqua, turtlehead. The yellow bearding is the stamp of this southern American swamp beauty.

MEDICINE

It is important to know that all parts of all of the elderberry species are poisonous, except for the flowers and fruit of *S. canadensis*, the American elderberry; *S. nigra*, the common elderberry; and *S. glauca* (syn. *S. caerulea*), the blue elderberry. Even the fruits of these species should be cooked before eating. The poisons of the elderberry are cyanide-based. In the case of the fruit, these cyanides are destroyed by cooking. Some people, especially some children, react to the raw fruit with a mild feeling of nausea.

The *Sambucus* species that are commonly used for medicines are the same three: *S. nigra*, the common elderberry; *S. canadensis*, the American elderberry; and *S. glauca*, the blue elderberry. All parts of these plants are used. The principal medicines of the *Sambucus* species are volatile oils comprised mainly of triterpenes. There are also fatty acids and rhamnose sugars, rutin and quercitrin, various alkaloids, some still not identified, anthocyanins, vitamin C, fats, resins, tannins, and mucilage. All of these chemicals are found in varying amounts in the fresh and dried flowers, flower buds, ripe fruit, leaves, root bark, and stem pith, which is the central soft, spongy region in the center of the elder branches.

The women of the Meskwaki aboriginal peoples used *Sambucus canadensis* in difficult childbirth. A tea was made from the bark that was taken to help uterine contractions and to shorten delivery time. This tea was used in a similar fashion to the present use of the pituitary hormone, oxytocin.

The mountain peoples of the southern Appalachians have, for several hundred years, used a tea of elder flowers as a tonic for the nerves. Flower heads of *S. canadensis* were collected. They were steeped in boiling water for two to three minutes. The water was strained. It was then sweetened with honey or sugar according to taste.

The Seneca peoples had a treatment for measles using *S. canadensis*, the American elderberry. The bark was scraped upward from a couple of young, green shoot suckers. This was steeped in 2 quarts (1.9 l) of well water for thirty minutes. A cupful was taken daily. This treatment probably contained one of the lectins called phytohemaglutin, which is presently being used as a molecular tool in cancer research.

The Onondaga aboriginal peoples had an interesting treatment for headache that parallels conventional migraine treatments presently in use. The bark is scraped from *S. canadensis*, the American elderberry. The piece is made into a poultice with warm water. It is immediately applied to the forehead. The warmth speeds up the entry of the rutin chemicals from the bark. These are highly water soluble. They enter directly through the skin and into the circulation for pain relief.

The Seneca also used *S. canadensis* for the treatment of heart disease. Young green shoots were harvested prior to July, before the plant became too toxic. Two handfuls of pith sponge were released from the central stem. This was placed into 3 quarts (2.8 l) of lukewarm water for not longer than three hours. The pith was strained off. The resulting water was drunk very gradually over time.

Both the Seneca and Cayuga aboriginal peoples used *S. canadensis*, the American elderberry, in an interesting management of their premature and newborn babies. The blossoms were collected. They were steeped in lukewarm water. The water was strained. Nets made from human hair were used for this type of separation. The baby was sponged with this warm water except for the eyes, ears, and nose. The mother's milk was reserved for these delicate areas. This water would contain rutin, quercetin, and isoquercetin, all of which are capillary protectants acting as a gentle skin tonic to help the baby's peripheral circulation.

In British Columbia, the aboriginal peoples, the Haida, Saanich, and Cowichan, used *S. glauca*, the blue elderberry, which is a 50-foot (15 m) tree, as a laxative. The bark was scraped upward and a decoction was made using well water. This was drunk as needed. They also used *S. racemosa*, the European red elderberry, in a surface treatment for rheumatic pains.

In England, the *S. nigra* was used variously with either peppermint, *Mentha × piperita,* or yarrow, *Achillea millefolium,* in the treatment of colds, flus, and nasal catarrh. A tea was also used as a gargle.

A wine made from *S. nigra,* the common elderberry, was used as a tonic in the Celtic culture for the old and the sick. It was thought to have an effect on the function of the brain. The vasoactive component of the flavonoids is thought to have a positive effect on Parkinson's disease and Alzheimer's.

The fruits of *S. nigra,* the common elderberry; *S. canadensis,* the American elderberry; and *S. glauca,* blue elderberry; are all high in a sugar complex called 3-rhamnoglucoside. This sugar has a remarkable effect on the eye. It rapidly helps the sight to adapt to light after dark and also to darkness after light. It is also of considerable aid to the aging eye for this reason.

The ancient Egyptians placed a high value on *Sambucus* as a cosmetic. The primary effect of *Sambucus* extracts is on the skin for rejuvenation and revitalization. This happens by means of a protective function on the capillaries that supply blood to the skin. Consequently, flowers of *S. nigra,* the common elderberry, were used for whole body baths and for removing aging of the skin around eyes by means of eye baths.

The elder flower water when used as a hair rinse has an astringent action that has the effect of highlighting the natural tints of the hair. The Romans used the anthocyanin pigments of the ripe fruit as a dark tint for graying hair.

Fruit juices made from the ripe berries of *S. nigra,* S. *canadensis,* or *S. glauca* are used in detoxification programs to help rid the body of cigarette smoking and drugs.

ECOFUNCTION

In North America and in Europe, *Sambucus,* elderberries, are always found near bodies of water such as oceans, lakes, rivers, streams, bogs, and damp areas. They act as significant staging grounds for the fall feeding of migratory bird populations, both north and south. An ample food supply directly affects a bird's ability to successfully reproduce in the spring. Almost thirty species of songbirds use both black- and red-fruited species of *Sambucus.* In particular, robins, mockingbirds, brown thrashers, and catbirds seek them out. Squirrels, chipmunks, rabbits, raccoons, skunks, foxes, opossums, and bears also take their fill. Game birds such as ruffed grouse, bobwhite, quail, mourning doves, and wild turkey have a great fondness for the fruit.

Nymphaea odorata, fragrant water lily. A pink form of this is found in great abundance in shallow pools on the north shore of Prince Edward Island, where it is the pink companion to a host of elderberries, *Sambucus.* Water lilies were used as a medicine to ward off ghosts by the aboriginal peoples.

In some years, when the summer has been humid, a yeast bloom can be seen on the surface of elder fruits. This yeast initiates fermentation. With the sugar changes in the pulp of the fruit and the formation of alcohol, the songbirds strip the elder as they also tackle the wild grape, *Vitis labrusca,* in competition with the game birds. These flying revelers quite often wreak havoc on windows.

All of the *Sambucus* species have their own antifeeding devices for mammals, which is the presence of a deadly systemic pesticide. The compound is a cyanide-containing glucoside called sambunigrin. This biochemical protects the more northern of the *Sambucus* species especially, since its suckers and leaves are so near ground level available for grazing. Snapping a twig or

Cephalanthus occidentalis, buttonbush, another shrub often found in the company of *Sambucus*. It flowers in July and August.

crushing a leaf releases the cyanide gas into the air, which is a truly nasty smell. When the cyanide is liberated, it quickly picks up hydrogen from the air and forms the deadly hydrogen cyanide or prussic acid that is so toxic. In fact there are few naturally occurring chemicals that can compare to prussic acid as a killing agent. Death is rapid by means of oxygen deficiency due to cyanmethaemoglobaemia.

The flowers of the *Sambucus* species produce a highly volatile terpene complex that is a chemical signal to great hordes of beneficial insects. These tiny insects like gnats and midges and flies like dance flies, which are predaceous, attend to the perfect flowers in large numbers as long as there are daylight hours. These insects feed on a nectar-resin complex produced by the ovarian area of the diminutive flowers. The elder is not usually visited by the larger insects such as honeybees, bees, and wasps in North America or in Europe.

The golden cultivars *S. canadensis* 'Aurea', *S. racemosa* 'Aurea', and *S. r.* 'Sutherland Gold' act as host species for the common spring azure butterfly, *Celastrina lodon,* and its various companions, for example, *C. ebenina*, the sooty azure. Their ova are deposited on the golden foliate, which acts as a food source for the developing caterpillars, who probably ingest cyanide for avian protection.

BIOPLAN

All of the *Sambucus* species, both red and black, should be used as environmental plantings in riparian areas, especially in the Maritime provinces where the moderating effect of the ocean helps to ensure a heavy crop of fruit. They can be planted in both public and private parks as feeding species for wildlife.

Sambucus species should not be planted around day cares, nurseries, and junior schools. This is because of the temptation to children to make pea shooters and whistles out of the wood. The placing of the freshly worked wood in the mouths of very young children could be toxic. These species should be avoided in campsites for the same reason.

Growing *S. nigra,* the common elder, and *S. canadensis,* the American elder, on the east coast and growing *S. glauca,* the blue elderberry, on the west coast could be an ideal cottage industry. These fruits grow particularly well near bodies of water. They have a potentially larger market than the grape because of the greater number of products that can be made from them. The fruits and flowers can be harvested for the fresh market or they can be dried. The flowers produce a sparkling drink that has a unique flavor of the oriental lichee fruit, *Litchi chinensis*. The fruits can be fermented into a wine or into a wonderful anise-flavored liqueur. Cold beverages can be made from the heat-processed fruit juice, which can be mixed with other juices for the beverage market and for medical detoxification programs. A lighter tonic wine could also be marketed for seniors as a health drink.

Elderberry products such as relishes, jams, teas, jellies, sauces, fruit rolls, leathers, curds, and cheeses could be part of a smaller cottage industry. Both elderberry flower and fruit concentrates could be marketed with the dairy and soya industries as ice creams, yogurts, cottage cheese, and cream cheese.

Based on its antiaging flavonoid content of quercetin, kaempferol, isoquercetin, rutin, and astragalin, the flowers of *S. nigra, S. canadensis,* and *S. glauca* will continue to be in demand in increasing amounts for the cosmetic industry. The fruity fragrances for shampoos, body lotions, skin rubs, skin cream, and general body care have never been fully exploited. Cash plantings of these *Sambucus* species and flower harvesting in June and July would put money into the pockets of a declining farming community.

DESIGN

For a large garden, the elders make an excellent backdrop. The deep green leaves are pinnate and alternating, blending in well with most plants. In June and into July they shimmer with a load of attractive, white, flat-topped panicles. These tilt forward attractively with morning dew or with rain. Then later on in the fall, they become a circus for songbirds.

Sambucus Species and Cultivars of Merit

SCIENTIFIC NAME	COMMON NAME	ZONES
S. callicarpa	Pacific Coast elder	5–9
S. canadensis	American elderberry	3–9
S. c. 'Aurea'	Golden elderberry	3–9
S. glauca (syn. *S. caerulea*)	Blue elderberry	6–9
S. nigra	Common elderberry	3–9
S. n. 'Fructuluteo'	Yellow-berried elder	3–9
S. n. 'Guincho Purple'	Purple elder	3–9
S. n. 'Aureomarginata'	Variegated elder	3–9
S. n. 'Pulverulenta'	Striped elder	3–9
S. racemosa	European red elderberry	3–9
S. r. 'Sutherland Gold'	Golden elder	3–9
S. r. 'Tenuifolia'	Fern-leafed elder	3–9

White *Chelone glabra*, turtlehead, found with *Sambucus*, elderberry, in the northern portion of the American continent. But in the south the magnificent pink *Chelone lyonii* and *Chelone obliqua* are found. The latter dips into Florida. Interestingly, these were used as antiwitchcraft medicines by the aboriginal peoples.

Generally speaking, for the flower garden, the *S. racemosa,* the European red elderberry, and *S. callicarpa,* the Pacific Coast elder, are more attractive plants. The trees have a more formed, blocky shape and with their fall swag of brilliant scarlet fruit are most attractive. The red species are also attractive to songbirds. Because their crop is so heavy, the winter birds can be fed too.

For the northern and colder gardens in zones 2–5, there are a number of interesting cultivars that could stand on their own as specimen plantings. There is *S. racemosa* 'Tenuifolia', which is of Russian origin and is extremely hardy. It is an excellent substitute for the heat-loving Japanese maples, *Acer palmatum.* This *Sambucus* cultivar makes a low, mounding cascade of arching and finely divided, delicate foliage.

This species is very useful in a large rockery. There is also a golden-fruited cultivar, *S. nigra* 'Fructuluteo', and there are two variegated cultivars, *S. n.* 'Pulverulenta' and *S. n.* 'Aureomarginata'. In addition, an unusual purple-leaved *S. r.* 'Guincho Purple' with wonderful dark leaves may find its place in a garden.

As part of the kitchen garden or in a damp corner of the fruit orchard, the American elderberry, *S. canadensis,* has a number of large-fruited cultivars. These are *S. c.* 'Johns', *S. c.* 'Adams', *S. c.* 'York', and *S. c.* 'Kent'. They were developed in Nova Scotia. These cultivars set huge heads of large fruit, which are readily harvested. They also seem to increase their fruit set if any one or all of these cultivars are cross-pollinated by a local, wild *S. canadensis* species, placed within flying distance of them.

Antennaria neodioica, smaller pussytoes, an American companion to *Sassafras albidum.* This one-hundred-year-old male plant colony is rare and carries antivenin biochemicals.

SASSAFRAS

Lauraceae

Zones 5–10

Sassafras albidum showing the mittenlike foliage. This is a herbarium specimen.

THE GLOBAL GARDEN

The fragrant tree *Sassafras albidum* is also known as *Sassafras variifolium*. This tree has been part of the North American landscape since it was the giant supercontinent called Pangaea. This is estimated to have been about 150 million years ago. When this land mass split up, some *Sassafras* slid with the divide, giving rise to a pack of new Asiatic species. *S. tzumu* resides in mainland China today, as does *S. vandaicuse* in Taiwan. The strange thing is that despite their fracture in time the three sisters still have an identical taste for deep moist sand a little on the acid side. All of the species also live to a ripe old age, each spanning up to a millennium. And, because of their ancient geological history, it can be quite safely said that the North American dinosaur lunched on this spicy tree, which grew so well at that time.

The sassafras has learned the trick of adaptation. Basically it is a tree of the tropical forests that has moved into the semitropics and has squeezed itself as far north as Canada. The largest trees are to be found in the moist heat of the Tennessee River valley. As the tree adapts itself to the cold, it also reduces in size until it grows as a smaller version of itself in Canada with the suckering characteristics of a large shrub. In Canada, the sassafras survives in the margins. In such a species is born the necessity of invention for life, and it is, therefore, more interesting from a biochemical point of view. As global warming proceeds,

Sassafras albidum var. 'Molle', also known as red sassafras or downy sassafras because of the glandular hairs on the leaves. This plant is extremely rare in the wild. It should be protected wherever it grows because of its medicines. This is a herbarium specimen.

Sassafras will be like other evolutionarily regressive species such as the lowly scouring rush of farmers' fields, the *Equisetum*. This diminutive once pushed a 100-foot (30 m) mop of growth into the dinosaur's mouth. It now is also beginning to have a heyday with acidification of soils, increased carbon dioxide, and heat. Both species, the *Equisetum* and the *Sassafras*, will rise again in the long tide of millennial change.

The word *sassafras* comes from the aboriginal peoples, the Narragansett, whose home was Long Island Sound. They had named this tree the *Sasauaka-pamuch*. The sassafras had been to them a source of medicine, as it had been to so many eastern tribes since time immemorial. So, in 1564, when the traveling physician Nicholas Bautista de Monardes from Spain saw this age-old wonder of North America with its panacea of cures, he fully understood what he had found. In a very short time a massive industry was set up. It centered around the harvesting, distribution, and transatlantic sale of *Sassafras* into the Old World of Europe, which was so desperate for new medicines.

Sassafras became the first pharmaceutical to be mass transported from America to Europe where it became a "rave" drug. Like many cure-alls, it suffered the fate of ignorance over the centuries for . . . small, in the molecular sense of enough, is still beautiful, but more will often kill. . . . It is like the medicinal drug morphine in this respect. A small amount of this drug is therapeutic and a large amount will kill. The fact that it will kill as an overdose does not make its therapeutic use any less effective.

ORGANIC CARE

Sassafras likes to grow in deep, rich soil of rolling hills with the sandy, well-drained texture of open, porous ground. While the tree does like soil on the acidic side, large trees have been known to grow on limestone karst in the southern states. These trees bear heavy crops of fruit.

Sassafras can be propagated by seed. The tree can also be vegetatively propagated by root or root sucker cuttings. The fruit clusters are produced on terminal twigs in the fall. Since there are male and female trees, not every tree will bear a crop of berries. Sassafras berries consist of a stone surrounded by fleshy tissue, as in a peach or apricot. Such fruit is technically known as a drupe. In the case of the sassafras this drupe sits in a red cup with a long red stem. When the drupe has changed color from green to blue-black, it is ripe. The fruit can then be quite easily either shaken or knocked off the tree and collected for planting.

Once the gardener has removed the fleshy pulp and washed the seeds in lukewarm water, there are two options open for propagation. They can be fall planted or stored and planted out in the spring. In any case, for short-term storage, they are put into plastic and into the crisper compartment of a refrigerator.

The embryos of most sassafras seeds exhibit a strong dormancy that is broken only by being placed into damp sand for 120 days at 41°F (5°C). The seed-sand mixture should be put into plastic, tied and stored in the crisper compartment of the refrigerator until spring has arrived.

Sassafras seeds should then be planted out after the last frost in rows 14 inches (35 cm) apart. They should be lightly covered with soil to ½ inch (1.2 cm). The soil should be well tamped. The rows are then covered with 1 inch (2.5 cm) of sawdust. The whole seedbed gets a final cover of Reemay that is rolled back in June.

Sassafras seeds may, at times, begin to germinate immediately in the fall. This can present a problem to gardeners in zones 5–7. So the trick used is to keep the seeds fully dormant until spring or to delay planting the seeds in the fall until the gardener is fairly sure that cold weather is on its way.

Apart from fire, which will kill young trees and seriously damage older trees, the sassafras is visited by two kinds of branch cankers. Pruning cankered branches back to sound wood solves the problem. This tree has the genetic chemical makeup of many species that can live for a long time. They are almost immune to fungal diseases. But they are susceptible to Japanese beetle infestations, especially in recent years in Canada, where this introduced beetle is getting a toehold. For the gardener, the damage is worse in the suburban garden where the soil has recently been moved. Spring patches of grass that are killed and can be scuffed free by hand are the first warnings of Japanese beetles getting out of hand. A flashlight and an evening inspection can readily find the beetles, which can be thrown into a small container of soapy water to kill them. The insecticide called Doom (a trademarked name) can also be used. This spreads the milky spore disease of Japanese beetles and aids in their natural control.

MEDICINE

Sassafras albidum is closely related to both the cinnamon and the camphor tree. These are tropical trees that have been mined for their medical drugs for decades. As with these trees, the calling card of the sassafras is aromatic fragrance. This fragrance is produced by a host of related aromatic hydrocarbons, most of which are volatile. The fact that they are volatile means that they have easy entry into the body, which they affect in an extraordinary way. They produce a feeling of well-being. Perhaps it could be said that an acute feeling of well-being is the true calling card of the sassafras tree.

The sassafras produces biochemicals that are related to myrrh, a gum resin of the *Commiphora abyssinica* tree found growing in Nubia. Myrrh was one of the legendary spices given to the baby Jesus as a birth gift from the three wise men of the East.

Jesus, Mary, and Joseph used this myrrh for the same reasons that the aboriginal peoples were using *Sassafras* at the same time in history. An aromatic oil may be extracted from all parts of the sassafras tree. This is known as oil of sassafras. This oil consists of about 80 percent safrole. There are additional compounds such as myristicin and related compounds, asarone isomers, phellandrene, and pinene isomers. Wax and mucilage are also found. Safrole itself is a benzine ether called 4-allyl-1,2-methylenedioxybenzene. This is also part of the familiar flavor of cocoa, star anise, nutmeg, black pepper, mace, and cinnamon, for it occurs in minute amounts in these household spices.

Lycopodium annotinum. Delicate club mosses like this make a fine ground cover for *Sassafras albidum* in a woodland setting. *Lycopodium*, like the *Sassafras*, was a more dominant species in the wide arena of previous global warmings.

Safrole flavored the famous North American drink called root beer. The benzene portion of the safrole molecule has been found to be carcinogenic when ingested in large amounts. It has also been found to damage the kidneys and liver, again, when ingested in large amounts. So in the 1960s, safrole was taken out of root beer and a harmless substitute added instead.

Sassafras was a very important plant for the aboriginal peoples of North America. It was used as a food enhancer, as a medicine, in rituals, and for such necessities as canoes for water transport because the wood is rot-resistant. Some of the medicinal uses ranged from the treatment of high blood pressure including nosebleeds, to an active antihistamine, which acted as a blood thinner used much as heparin is used today.

The Seneca people handled eye inflammations and cataracts in a very interesting way. The fresh, young twig growth contains a mucilage. This mucilaginous liquid was milked by finger

The native cormous *Delphinium menziesii*, larkspur, from the northern California coast. It makes a designer's delight with *Sassafras albidum* in any garden.

squeezing. The resulting liquid is sterile and aseptic. This was dropped into the eye as a soothing, anti-inflammatory eye lotion.

The Choctaw aboriginal peoples used *Sassafras albidum* and a locally occurring variant known as *S. a.* var. 'Molle', or red Sassafras, as a tonic tea. Before March the bark can be slipped off the roots. This was cut into strips and dried. Segments 3 inches (8 cm) long were put into boiling well water. This was steeped for 4–5 minutes. The tea was sometimes sweetened.

The mountain people of the southern United States use sassafras tea stirred with a stick of spicebush, *Lindera benzoin,* or sweet birch, *Betula lenta.* The tea was made of flowers, twigs, or roots. It was drunk hot or cold on ice. A hard candy and a bottled sassafras jelly was also made. It was sometimes used on February 14 with golden seal, *Hydrastis canadensis,* and sweet birch, *Betula lenta,* as a potent spring tonic to strengthen the immune system and the heart.

It is as a chew stick that the sassafras comes into its own. Chew sticks have been used all over the world in dental care. They are the common denominator for good teeth and a healthy gum line. Chew sticks are part of our ancient heritage. They are chosen from some local medicinal tree that is either antibacterial, fungicidal, or high in fluorine. In all cases the stick is chewed and the gum line is massaged. This action seems to remove dental plaque and produce excellent dental hygiene. This custom is still practiced by the Iroquois peoples, who remove terminal, pencil-sized twigs of the *Sassafras* and use them as chew sticks.

In modern dentistry, sassafras is used in combination with *Capsicum,* pepper, and *Humulus,* hops, in commercial dental poultices to which benzocaine and hydroxyquinoline sulfate have been added. This mixture disinfects root canals in dental surgery. Sassafras is also added to some modern toothpastes in a combination with other mouth-freshening flavorings.

Sassafras, in large quantities, is said to have hallucinogenic properties. This is probably because of an increase in the increments of an analgesic narcotic called myrophine. In commerce, sassafras has been used in soaps, perfumes, and shaving creams. It is used as an artificial flavor such as cherry and vanilla, as a wetting agent in the textile industry, as a lubricant, and as an emollient for cold creams and in anodized coatings for aluminum. Some sassafras extracts are used in the manufacture of lenses and in glass prisms.

Using cream of tartar as a mordant, pioneers obtained a rose-tan dye in wool and a gray to dark rose-tan in cotton using the root and bark of the sassafras tree.

Even to this day, sassafras is used as a revitalizing spring tonic. A tea is made of the young spring roots. It is also part of Cajun cooking. A spice powder, called "gumbo filet," is made from the first growth of spring leaves, which are dried and then crushed. This is used to thicken soups and in a stew called jambalaya in the Creole and Cajun cooking of Louisiana.

ECOFUNCTION

Sassafras is a feeding tree. Birds, both large and small, use it, as do insects and mammals. The dark sassafras berry, produced after the tree is about ten years old, is hungrily sought after by at least 14 species of songbirds. The berry, seated in its scarlet throne, is readily edible to bluebirds, the thrushes, and mockingbirds and to the larger birds, wild turkeys, ruffed grouse, and bobwhite quail. As wild turkeys are being reintroduced into Canada after many years, their success is often dependent on fruit-bearing trees such as the sassafras.

The winter twigs and leaves provide browse for deer herds. Larger mammals such as the raccoon, the squirrel, and the black bear make use of the fruit.

The mature leaves of the sassafras are the host food for two beautiful swallowtail butterflies, the spicebush, *Pterourus troilus,* and the palamedes, *P. palamedes*. The latter is the signature butterfly of the great southern swamps. The spicebush is becoming more common in Canada as global warming proceeds. A moth known as the Promethea moth also uses sassafras leaves as a feeding host. Since these butterflies are highly colored and seem to be loathed by predators, there is little doubt that the chemical arsenal of the sassafras is converted to their use to deter predation.

In the early spring the fragrant, yellow flower clusters produce an abundant nectar supply for honeybees, among other insects.

The aromatic sassafras is one of the first trees to invade abandoned fields in warmer climates. The roots produce safrole-containing allelochemicals that eliminate sensitive competition, and thus the ground is prepared for the incoming forest. The aerial parts of the sassafras produces a fragrance that carries as part of its chemistry a volatile hydrocarbon called β-asarone. This unique isomer has the ability to act as a chemosterilant on some insects. Thus the sassafras acts as an insecticidal aerosol maintaining the health of the young forest.

Sassafras timber has a number of unique qualities. It is fungus resistant, and it is moth and louse repellent. Sassafras, because it did not decay in dampness, was used as lintel wood for doors and windows. It was also used for dugout canoes and fencing.

The wood was used to build chicken houses as a natural remedy to keep high louse populations from damaging chickens. Household chests were made from sassafras to keep moths away from stored clothes. Often slave cabins were built from it to control lice on the unfortunate inhabitants.

It would be logical to suppose that the mycorrhizal growth associated with the roots of the sassafras is unique because there are no known fungi of the higher orders associated with this tree.

BIOPLAN

Sassafras albidum is a tree of the ancient virgin forests of eastern North America. It is a tree whose boundaries ebb and flow with the tides of global climate change, becoming more dominant and strong in parching heat and receding with the cold. Sassafras can live as a canopy tree or as an understory tree biding its time in the shade waiting for dominance. Its thousand-year life span is a legacy of survival. Any serious reforestation project should include *Sassafras albidum* for its ability to be flexible within the

Viburnum trilobum, American highbush cranberry. It adds an exotic touch of red to *Sassafras* in the fall.

long arm of change and because it is a feeding tree that produces aerosol chemicals that protect the surrounding forest from insect infestations.

The sassafras is an ideal small tree for the northern suburban garden. A male and female tree should be planted to produce a crop of berries. All parts of this tree are fragrant, from its early spring flowering to the spectacular fruiting in the fall. The aromatic fragrance produces a corridor of health-giving airways around and into the home. It invites both swallowtail butterflies and will guarantee the presence of songbirds in the garden that in turn will reduce the need for pesticides because a songbird's feeding pattern on noxious insects and weed seeds deletes the need for spraying. Thus the entire household, even the family dog, is more healthy.

Hospitals, retirement homes, seniors' apartments, and walk-in clinics would benefit from the planting of *Sassafras albidum* or the use of sassafras mulch. The infirm and particularly older people with slower metabolisms because of chronic illness or reduced movement would benefit from the aromatic airways produced by these trees. The airways are at once a stimulant and diaphoretic. This means they help to increase sweating and thus aid in ridding the body of toxins. It is a stimulant because some of the aromatic hydrocarbons act to open up the arteries of the brain, as a cerebral vasodilator. This, of course, induces clearer thinking. The fragrance also has an antihistamine fraction that helps reduce the effect of allergens. The chemicals in the fragrance also increase the quality of breathing. This in turn improves blood flow. The blood is made "'thinner" by a decrease in viscosity. All of these functions are intensely beneficial to the sick and old.

An innovative sweat lodge could be designed for the sick and convalescent for very little money. In the past, wood shavings from the sassafras were used in combination with two other species as decoctions to induce sweating. They are *Guaiacum sanctum,* lignum-vitae, growing in southern Florida, and the lily, *Smilax ornata,* also known as greenbrier. The root of *S. ornata* is used. *Guaiacum* is a traditional medicine of the tropics used for the maintenance of the urogenital system. As well as being expectorant and antitussive, it acts as a muscle relaxant. There is a side reaction of anti-inflammation and antiulceration. *Smilax,* on the other hand, is a tonic with an action similar to ginseng. A combination of shavings of these three species could be used as a mulch in a sitting or resting area outdoors. If the design were such that sunshine were to fall directly on the mulch, then the resin component of the mulch would slowly sublime over the summer months. The sun's action would release beneficial aerosols around the seated people, enabling them to inhale the medicine from the three species in trace amounts, and, as a result feel, invigorated.

DESIGN

As a tree of great beauty, *Sassafras albidum* has long been appreciated in North America. This is seen by its host of common

Sassafras Species and Cultivar of Merit

SCIENTIFIC NAME	COMMON NAME	ZONES
Sassafras albidum	Sassafras	4–9
S. a. var. *molle*	Red sassafras, downy sassafras	4–9

names, all of which are generated by country living with this tree. Tea tree, mitten tree, green stick, smelling stick, ague tree, cinnamon tree, red sassafras, saloop, saxifrax, and sassafras are but a few of them. All of these common names addressed the world's largest sassafras in Owensboro, Kentucky. This *grande dame* is around three hundred years old with a circumference of 21 feet (6.4 m) and a diameter of 6.6 feet (2 m).

In design the form of the mature sassafras is square, with long horizontal branches leaving the trunk at right angles. The terminal branches break into an irregular pattern of green, fragrant twiglets that look like bunches of staghorns. Each terminal is attractive in itself, carrying plump, green, scented buds enfolding male or female flower embryos, which will, in turn, be both fragrant and yellow.

The tree carries three types of leaves, a lance form and a right and left mitten form. These have spectacular fall colorings of orange, gold, and flame. They drop, leaving a long, dashed, bud scar, which is a key toward winter identification of this tree.

In Canada, where the tree assumes a smaller stature, the vermilion fruit cups are more easily incorporated into the overall garden design. Sassafras can be planted with *Euonymus elata,* burning bush, or *Viburnum* ssp. for spectacular, red fall coloring. Either *V. plicatum*, *V. alnifolium*, *V. cassinoides,* or *V. lentago* can be successfully used as balancing shrubs. For a more exotic touch of red, one of the many late-flowering clematis vines can be induced to wrap around the sassafras, such as the *Clematis viticella* 'Madame Julia Correvon', with its red, semi-campanulate, stunning flowers in September.

In general, in the north temperate gardens in the fall, there is a bronzing of the foliage and twigs that takes place prior to the really cold weather. The reds of sassafras act as a catalyst for these subtle colors, calling them to attention.

Beautifully mounted flowers of *Sassafras albidum.* These were collected in southern Ontario. Few people have seen these flowers. They live on the margins of life.

The extraordinarily drought-resistant native low grasses that were part of the open spaces in virgin cedar forests. Deer show a propensity for these areas for bedding down.

Thuja occidentalis

CEDAR

Cupressaceae

Zones 2–9

THE GLOBAL GARDEN

In the global garden the eastern white cedar grows naturally throughout eastern North America from Nova Scotia, with a sprinkling in Prince Edward Island, to the southern shore of Lake Winnipeg southward to the Carolinas and as far north as James Bay. The white cedar grows to its maximum potential around the Great Lakes and trails off to a shrub in the Maritimes and the elevations of the southern Appalachians.

Few mature specimens of white cedar are to be found on this continent. A harvest of 95 million board feet of white cedar in 1899 from the Great Lakes region bears testimony to its being part of the wildwood or virgin forest of North America. Because it takes a minimum of 200 years of growth to produce quality boards, this places these trees as seedlings in the fifteenth and sixteenth centuries at a minimum. White cedar was unknown in Europe at these times and represented a new species for the lumber barons of the day. White cedar is and was a gregarious species of tree; it seldom occurs as a single specimen, usually preferring its own company to that of other trees. History, the little of it recorded regarding forests, and the artifacts that remain are ample evidence for the importance of white cedar as part of the wildwood.

There are mature, albeit, dwarf white cedars in Niagara and in parts of Quebec that are over 700 years old. A mature white cedar planted in good soil is a sight to behold. One of the Lords Lansdowne planted such a tree on his estate in southern Ireland less than two centuries ago. This tree, in its maritime, semitropical setting, takes on the distinct character of the giant sequoia, *Sequoiadendron giganteum,* of the west coast and is probably one of the world's largest trees. They are both of the same family, the Cupressaceae. We know the full potential of the sequoia because we have a few protected plantings remaining. This is not so for the eastern white cedar.

Another relative holds the fort in the rain forests of the west coast of North America, the giant western red cedar or *Thuja plicata.* This species and the eastern white cedar are the two species of cedar native to North America. Six species were known globally until 1941. A seventh existed only in fossil records until the astonishing discovery of the dawn redwood, *Metasequoia glyptostroboides,* in China, known by the villagers as the old "water fir," growing by an equally old temple. Seeds were sent to the Arnold Arboretum at Harvard University and the dawn redwood was on its way to popularity after a three-million-year wait on the sidelines.

In Canada, especially eastern Canada, the white cedar is so much a part of the visual landscape that its importance is largely overlooked. In farmers' fields, cedar rail fences snake into patterns of complexity that are unique in character from district to district. Cedar is and was the building material of choice for log homes both ancient and new. It is found on the roofs of houses and barns as cedar shakes. It forms the ribs of the old-style cedar and bark canoes and controls the massive flow of the canal systems' waterworks as stop logs. Bowls and spoons are commonly made from cedar. Even the leftover cedar slab from sawmills fires the reduction kilns for local pottery. Bakers built fast, hot fires necessary for the crunchy crust of breads in their wood stoves of split cedar logs.

The white cedar comes into its own as the perfect urban dweller. Being an evergreen, it makes an excellent hedge. In fact in the colder zones of 2, 3, and 4 in Canada and the United States, cedar is the only really reliable, living, evergreen hedge.

Thuja occidentalis, eastern white cedar, with a stool of sixteen feet. The virgin tree was cut about 150 years ago. The living roots continued to grow to produce a multitrunked bole. Such areas of first-cut forest are rare now in North America.

By contrast to the cedar hedges of England, the white cedar hedge in North America requires a minimum of pruning. *Thuja orientalis,* the cedar commonly grown in England, is oriental and is the fast-growing substitute for the glorious yew hedges of the English "grande" garden. The many cultivars of white cedar make it the darling of horticulturalists and gardeners alike. Even the smallest city balcony has a space for a miniature cedar. In a cold winter's landscape cedars make the framework of a garden stay alive. For the country dweller with a little extra land, cedars can be planted in groves whose filtered light is an experience of extraordinary beauty.

Perhaps its other name, that of arborvitae, gives the real reason why the white cedar is so important. *Arborvitae* is a Latin-French word mixture meaning "tree of life." This naming comes from the casual observations of the early pioneers in eastern North America many centuries ago that wherever there were forests of white cedar, there was also a multitude of wildlife. In fact, a cedar forest teems with animal life. This is especially evident in winter when there is snow on the ground. The fascinating mixture of animal tracks found in the snow bears witness to the savory substances in the fronds of cedar.

ORGANIC CARE

White cedars love an alkaline soil. In particular, they like limestone bedding planes. This, alone, is the secret to successfully growing cedars.

Cedars are adaptable. They will grow in both wet and dry situations. In very dry habitats the cedars will quite often be hollow. Cedars respond to a fertile soil in an extraordinary manner with growth rates from 1–2 feet (30–60 cm) even in a dry year.

Cedars are usually transplanted when they are about 3 feet (90 cm) tall. The roots of a healthy cedar are red and firm. If this is not the case, then the tree has root decay, to which they are prone when young. Trees exhibiting this decay will not thrive. Broken or damaged roots should be trimmed to a clean cut.

When selecting trees for planting, the southern aspect of the tree should be marked, noted, and respected.

The planting hole should be 2 feet (60 cm) in diameter and 2 feet (60 cm) deep. Good-quality topsoil should be mixed with two shovels full (4 l) of sheep manure that is at least one year old. Phosphate in the form of steamed bonemeal (1 cup, .25 l) should also be added to the hole and mixed. A similar volume of ground limestone should be added and mixed. In a very dry area, 2 quarts (1.9 l) of dry peat moss should be added to the planting hole to carry the cedar through the first summer with a minimum of watering.

While planting, the tree should be held steady and perfectly upright in the hole. The roots should be spread out horizontally and should be carefully packed with soil. Tamping will have to be done gently because of the roots' horizontal and brittle nature. A catch basin is made with a collar of soil around the hole to collect and hold rainwater around the tree, which is initially well watered.

Quackgrass, *Agropyron repens,* can seriously impair a cedar's ability to thrive in the first five years. This is especially true of cedar hedges. Cedars for hedging are spaced about 1 foot (30 cm) apart. The cedars should be mulched by a 4-inch (10 cm) mulch of either rotten hay or straw. If this is not available, fresh leaf litter in the fall from any deciduous source should be raked to the tree or trees, making a circle of about 2 feet (60 cm) wide. The fresh tannic acid complexes of the leaves make an efficient natural herbicide that is nontoxic and whose residual effect will be approximately one year. This can be replaced if necessary. Subsequently, cedars should be side-dressed once every five years with 1 cup (.24 l) of dolomitic lime or 1 pint (.5 l) of extra dry woodash.

Dwarf cedar cultivars can be planted in large pots or barrels for patios, balconies, terrace gardens, or rock gardens. The containers must have good drainage. The soil mixtures should have a ratio of 20 parts good garden soil, 2 parts sheep manure, 1 part bonemeal, and 1 part dolomitic lime added and mixed. Then 6 parts dry peat moss should be added and mixed to counter des-

Cladonia. With their bright red fruiting bodies, these lichens use dead or decaying cedar as a carbon source.

iccation. The soil surface itself, after planting, should be mulched with round stones or bark chips for moisture conservation during the hot windy days of summer.

Because white cedar is an evergreen, it can suffer from water starvation during periods of extreme cold. In the colder zones of 2–5 when the soil is frozen for five to six months of the year, cedars should be thoroughly watered before the onset of the winter months. Cedars shut down photosynthesis at 50°F (10°C) and below. Watering should take place before dormancy sets in when it is observed that the foliage is beginning to turn green-brown in the early winter.

In urban and city areas, cedar hedges are sometimes protected from the chemical burns of calcium chloride—road salt mixtures that are spread on ice-covered roads. A wrap of a single layer of burlap can be used or a single layer of a protective spun polyester fabric such as Reemay. These fabrics have the advantage of being able to breathe during cold winter months.

Cedars can be pruned for topiary or for hedging in spring. This is usually done a few weeks before growth starts. In the warmer parts of North America for zones 6–9, this is usually February to March. For zones 4–5 a pruning in April is usual.

A stranger in paradise. Never think that the job of classification is over. This as yet unclassified species sits with *Osmunda regalis*, the royal fern, in a wet *Thuja occidentalis*, eastern white cedar woods in eastern Canada.

For zones 2–3 any time from April into early May when the temperatures begin to rise over 50°F (10°C).

The many cultivars of cedar are propagated professionally by means of softwood cuttings with the aid of auxins or plant hormones. They are then induced to growth by an elaborate misting system. For the average household, wild white cedar is more readily propagated by seed. The seeds fall in great numbers from December to January and can be hand collected. The cones can also be picked in October when they have turned a muddy brown. They are placed on newspaper in a warm, dry place. The cones open in 7 to 10 days. A few finger taps will release the seeds. Since white cedars have a high percentage of empty seeds, the sterile and fertile seeds should be separated. A cupped hand and a gentle blowing by mouth is all that is necessary. Fertile seeds will have a dark brown, plump eye in the center of the seed. These can be stored for up to five years, chilled, or at room temperature.

A higher rate of germination is found in fall planting for white cedar. Seeds are sewn at a depth of ¼ inch (.6 cm), with approximately 50 seeds per square foot. The seeds are then subjected to temperatures of 86°F (30°C) for 8 hours in light and 16 hours at 68°F (20°C) in darkness. The seeds will take approximately twenty-one days to break dormancy. The accuracy of these temperatures does not have to be absolute. It is the diurnal rhythm that is important to persuade the seed that conditions are ready for germination.

In its first year the seedling cedar will benefit greatly from half-day shade. This seedling will have two cotyledons or embryonic leaves and will produce a strange spindly leader from between these leaves that looks nothing like the mature cedar tree. The second year sees a lopsided fan of green leaves produced. From that point onward the seedling will grow rapidly and progress toward looking like its adult parent. It can be transplanted in the next year or so.

It is a fact of life for those of us who live in the countryside that deer are fond of cedar. In many cases, the finer the specimen, the greater the attraction. This has had the effect of reducing many a gardener to tears and the deer in question to venison pie. There is an answer to this problem, my unique solution. It works! Take a white plastic bag, the kind acquired in grocery stores across the continent. Lay it flat on a table. It will be seen that the bag has one handle on the left and another on the right. Cut the bag in half down the middle, giving you two equal halves, each one with a handle. Each half has now become a wind sock. Hang one or two from a cedar at 4–5 feet (1.2 m–1.5 m) in height. The wind sock has now been transformed into a danger signal for the browsing deer. The white fluttering bag is an alarm signal that is a repetition of their own tail alarm signal. A deer's tail has a white underpatch that, when the tail is lifted, is exposed in times of danger. This is a genetic imprint on all white-tailed deer. It is something they will not ignore. This is effective in both the orchard and the nut grove.

MEDICINE

Few households outside of the aboriginal peoples' and few laboratories outside of the multinational drug companies' realize the importance of the two *Thuja* tree species, *Thuja occidentalis*, eastern white cedar, and *Thuja plicata*, western red cedar. This is in stark contrast to the ancient Romans and Greeks, who both understood and highly valued cedars for the healing medicines this family of trees offered to mankind. These two civilizations must have acquired their knowledge from a previous civilization because by the time they arrived on the scene the cedar wildwood was almost gone from the Mediterranean area. This had the effect of increasing the market value of the medicines. We do know it is said that the infant Jesus in Bethlehem was offered a birth gift of a vial of incense from the cedar of Lebanon. This was a very valuable offering indeed, two thousand years ago.

The cypress or Cupressaceae family is world famous for its incense oils and potions. All members of this family are aromatic. This means that the trees manufacture aromatic oils as part of their daily routine of life. These oils are also called essential oils whose ingredients vary a little from one tree species to another in both degree and content. The essential oil from white cedar is called cedar leaf oil, which is constant to the entire white cedar family, including all of its large tribe of cultivars.

Studding the scaly leaves and small branchlets are tear-shaped glands just visible to the naked eye. These track the upper and underside of the fronds like tiny footprints. With the aid of a hand lens, the glands can be clearly seen. They appear to hold one small teardrop of light yellow essential oil that has an opaque appearance. Injury with a thumbnail to this gland releases a strong and pleasant odor of sage mixed with camphor and a touch of lavender. This is the essential oil of white cedar. It is part but not all of the tree's living pharmacy.

Commercially, this essential oil is extracted from fresh, green cedar boughs by a process called steam distillation. If a droplet of this oil is examined by another means called gas chromatography, the oil is separated into a number of very interesting chemicals. These are α-pinene, α-fenchene, camphene, sabinene, β-mycrene, limonene, α-fenchone, α-thujone, β-thujone, camphor, and bornyl acetate. Many of these chemicals are related. Then there are also other stray dog chemicals from other parts of the tree called thujic acid, pinitannic acid, and thujin.

Across time the ancient Greeks used cedar in their sacrificial fires, which they used to fumigate or cleanse the sacred area. The broad spectrum of antibiotics resident in the hardwood did an excellent job for the Greek priesthood. Simultaneously, or perhaps even a little earlier, the aboriginal peoples of North America were performing a similar task but to a different end. The heartwood of cedar was used to fumigate and strengthen the spirit of man in the sacred ceremony of the sweat lodge. White cedar heartwood logs were burned, releasing a highly potent broad spectrum antibiotic called thujic acid into the closed-air breathing environment of the lodge. Thujic acid also has a fungicidal action, so the cleansing action on man was absolutely complete.

Probably on the advice of the aboriginal peoples, the early settlers used white cedar as a besom sweep. Since this had been an ancient practice in the Celtic world, the idea was not a new one to them. The sweep of the fresh, green besom was used to rid the household of pests and vermin. The α-pinene, camphor, and limonene released are particularly potent antivermin compounds.

Externally only, the Iroquois used a mixture of white cedar and balsam fir, *Abies balsamea,* branchlets that were first pounded and then boiled for the treatment of sprains, bruises, and cuts. The area was bathed with the solution mixture, which was applied as warm as possible to aid in penetration of the active ingredients into the oils of the skin.

To counteract the more violent pains of severe sprains, cedar cones were powdered and one-fifth of this volume was added to the leaves and roots of a fern called common polypody, *Polypodium vulgare.* A poultice mixture was made with warm water or with warm milk and applied as hot as could be borne by the patient. Care had to be taken to ensure scalding did not occur with the milk poultice mixture.

In the treatment of rheumatism the young growing tips of cedar were harvested and boiled. Feet were then plunged into the cooling water, which was as warm as possible. The heat and the vapors were trapped around the legs using blankets. An antirheumatic ointment was also made using a paste of young leaves pounded into any animal fat. A modern substitute for this fat would be lanolin. This ointment was then used locally to ease pain.

For treatment of colds and flus, boiling water was poured over a bowl of fresh green branchlets. The vapors were inhaled with the use of a towel draped over the head. The expectorant action was immediate.

In casual living for people who suffer from any form of heart disease, relaxing, sitting, or strolling around white cedar is beneficial. This should be done on a warm summer's afternoon when the sun's rays have warmed the foliage and the temperatures are near 80°F (25°C). The trees will release fenchone and both α- and β-beta thujone into the immediate environment. These chemicals are strong cardiac muscle stimulants and will

Spiranthes cernua, nodding ladies' tresses, a fragrant, white orchid always found associated with *Thuja occidentalis* in wildwoods. It blooms in August in the northern portions of the continent. The flowering is delayed into the late fall as this species tracks southward.

Thuja occidentalis

It was quite a few years ago that I stood spellbound one sunny morning watching my neighbor tip a large ewe over and proceed to remove what seemed like a knitted pullover—four legged of course—from the beleaguered animal in the blink of an eye. He cast a shrubby eyebrow in my direction.

"They do this much faster in Australia, ye know."

I didn't.

"Nearly lost this bunch!" The shrub looked proudly down at his twenty north country Cheviots.

"Yes?"

"Snowstorm. D'-ye-know. The dirty thirties!"

"Oh!"

"An' money in m' pocket."

"Got 'em out of the cedar bush in the spring. Ate cedar all winter jes' like the deer!"

I was astonished.

"All 'em lived," the shrub twinkled on. "Had 40 lambs that year, too. An' money in m' pocket!"

help the heart in its pumping function, from which a considerable health benefit will arise.

White cedar also produces camphene and camphor, which are bronchodilators and improve the breathing of all those suffering from chronic lung problems.

Cedar bark that has been aged and dried for over one year makes an excellent vermifuge smoke for beekeeping activities. This was used by the early pioneers.

The lives of the famous Jacques Cartier and his crew were saved by being offered a tisane of white cedar as a treatment for their severe scurvy condition they suffered while on their travels due to an acute and prolonged lack of ascorbic acid, or vitamin C, in their daily fare. The tisane decoction was a cure offered by local aboriginal medicine for this disease.

ECOFUNCTION

The ecofunction of the white cedar, *Thuja occidentalis,* in eastern North America is profound. The DNA of cedar has learned to adapt to the harsh extremes of a continental climate with severely cold winters and hot, humid, breathless summers with an offering of coping chemicals for the survival of the plant and animal world around it.

During the months of winter, when starvation is a state that every animal must deal with sooner or later, the white cedar steps in to help. The evergreen boughs of cedar swept down to the snow, especially in a hard winter, produce a taste enhancer in their branchlets and fronds. The taste enhancer is a chemical called fectone. It makes the boughs of cedar extremely attractive as a food source for sheep, deer, rabbits, cottontails, hares, squirrels, and chipmunks, who in turn supply food for foxes, fishers, wolves, coyotes, lynx, and bobcats. The food line brings in weasels, martins, ermine, and sometimes ferrets.

It is no casual act of nature that the evergreen fronds are designed as perfect chemical applicators. In the sweeping, forward and backward movement over a person or animal as it passes through a cedar grove, the tear-shaped glands are forced into dispensing a chemical load on the passerby. These chemicals are all beneficial and even carry with them a dispensing compound called limonene for better dispersal. In all animals these chemicals get absorbed into the deoxy form of vitamin D on the skin, feathers, or fur and get ingested by the peening or preening process in sunlight, where vitamin D conversion takes place. Man takes his share of this treasury in the same biochemical way if he takes a walk in the woods or perhaps does a little gardening. The antibiotics, fungicides, bronchodilators, and heart muscle enhancers are metered out in minute parts per billion or parts per trillion doses. Enough, anyway, to keep health in balance.

Within a forest or garden the white cedar acts as a black box or energy trap. This is as important for avian or bird warming as it is for gardening. Cedars absorb a full 2 percent of the sun's solar energy to which they are exposed. This is divided up into 1 percent for green business in the chloroplast and 1 percent that is radiated out again as heat. This heat gain can crank up a cedar hedge one full climatic zone. In other words, a microclimate is produced and a zone 5 area becomes a zone 6 area in a garden. For those of us who are always trying to do the impossible in our gardens, this presents a very attractive proposition to increase the planting possibilities.

Bird populations make considerable use of cedars as a food source for seeds, as a heat sink in the forest and as a "safe house." Predation of songbirds is virtually impossible in white cedars.

Shredded cedar bark is an important nesting material for squirrels and particularly for the elusive flying squirrels.

In the summer when the temperatures soar to above 80°F (25°C) and the humidity index is very high, a form of solar distillation takes place around cedars. These trees become much more aromatic and intensely fragrant. The chemicals in the leaves are being converted into an aerosol form. These may form airborne allelochemicals that define the tree's territory, along with converting the trees into a health-giving spa.

BIOPLAN

In general terms, trees form an important part of any bioplan because of the shelter, feed, nesting, and perching they offer. The white cedar can be considered to be a medicinal tree and as such should be part of a bioplan for nurseries, day cares, and junior and senior high schools. Children of all races have one item in common as they grow into young adulthood. It is called early childhood sickness. It is nature's way of strengthening the immune system to produce a healthy adult. The white cedar offers considerable prophylaxis to the health of children. Cultivars of white cedar should be designed into the bioplan in such a way that they become tactile to the children at play. Walkways in a design should also be narrowed into a bottleneck area so that there is some brushing against or physical contact with the cedars themselves.

There is a secondary benefit to using white cedar in the bioplan of schools. All of the cultivars produce fungicidal compounds in the heartwood. Because of this there is very little spontaneous fungal growth in the form of poisonous mushrooms associated with these trees. The one fungus associated with cedars is a unique green mushroom called *Hygrophorus psittacinus*. It is not known if this species is edible. In addition, as sum-

History in a split-rail cedar fence. This delightful pattern of very old cedar comes from the virgin cedar wildwoods of the area.

mer temperatures climb, white cedars produce a fire-retardant compound, making the white cedar a relatively safe tree.

Bioplans for hospitals and buildings where air quality is impaired due to poor ventilation should design recreation areas involving the trapping of the sun after midday to sundown around plantings of white cedar trees and cultivars. The heat sink trap will force the cedars to vent their treasury of aerosol chemicals into the atmosphere. This will be greatly beneficial for the recovery of patients and the general well-being of the people who suffer from "sick-building syndrome."

It goes without saying that the white cedar should be part of the bioplan around an ordinary household. If there is limited space, the cedar should be the first tree of choice. Since there are so many cultivars to choose from, it becomes a matter of individual taste. The weeping *Thuja occidentalis* 'Pendula' adds particular beauty in addition to its other attributes to any garden.

Global warming is something the world will have to learn to live with for the next generation at least. In Europe there is hot debate about windbreaks, trees used to slow down the excess winds of increasingly dangerous storms. Shelter hedges of white cedar slow wind speeds to less than one-quarter of that above the canopy. Thus, wind speeds of 60 mph (100 kph) would be a mere 15 mph (25 kph) or less on the ground.

Monocropping of trees as forest reclamation is the antithesis to a forest bioplan. Trees have natural companions in a forest. They share a similar need for the same type of soil, and in some cases they share the same soil microbia. The companion trees for white cedar are white elm, *Ulmus americana,* yellow birch, *Betula alleghaniensis,* silver maple, *Acer saccharinum,* and black ash, *Fraxinus nigra.*

DESIGN

The white cedar, *Thuja occidentalis,* or the American arborvitae as it is called in Europe, grows to a tree of outstanding proportions, reaching heights at maturity of up to 60 feet (18 m) or more.

The red-brown bark of a 150-year-old tree is quite extraordinary in its textured open patterning. The trunks of such trees are very wide at ground level and gently taper to the growing tips. The foliage of an open grown tree is sweeping, studded with dainty cones. The overlapping scalelike leaves are borne in four ranks ending in flattened fanlike sprays. The spray tips have truly beautiful globose meristems that show a lemon-green color before they grow in spring. Both male and female cones called strobili are borne on the same tree. In early spring, some time in March or April, the male cones are a dull red color. Each cone is composed of up to 10 overlapping woody scales that are attached at the base. The cones mature the first year. The older cones are also found on the trees and only add visual interest. Cedars are bright green during the growing months and a change to a dun-green in winter. Many of the gold cultivars undergo this color change also. One cultivar, *Thuja occidentalis* 'Wareana', shows very little color change from summer to winter.

In the nursery trade there are some strikingly beautiful forms of white cedar. Over the years these have been mainly developed in Europe and are much loved there. Examples of cedars for rock gardens are *T. o.* 'Woodwardii' and its golden version, *T. o.* 'Woodwardii Aurea'. These are perfect green and golden globes and because of their density are not easily damaged by snow. These cedars are useful in zones 2–5, as are *T. o.* 'Holmstrup' and *T. o.* 'Holmstrup Yellow'. These cedars have a more squarish, dense appearance, and conical is not quite the correct description for them, but they have a splendid solidarity of form. They are as slow-growing as they are hardy. In the warmer zones 5–7, the famous *T. o.* 'Rheingold' is a showstopper with its old gold color in the winter garden. Over time this shrub becomes very large. The *T. o.* 'Globosa' and *T. o.* 'Golden Globe', with year-round golden foliage, too, are useful in zones 4–9.

T. o. 'Filiformis' and *T. o.* 'Fastigiata' are cultivars that for the most part are slow-growing and are narrow upright trees with no evidence of side branches. These cultivars make excellent hedges, especially where space is at a premium. They do not need pruning except a little tipping when they become too tall. When

Thuja and Cultivars of Merit

SCIENTIFIC NAME	COMMON NAME	ZONES
Thuja occidentalis	Eastern white cedar	3–9
T. o. 'Beaufort'	Beaufort cedar	3–9
T. o. 'Fastigiata'	Fastigiate cedar	3–9
T. o. 'Filiformis'	Filiformis cedar	3–9
T. o. 'Globosa'	Globose cedar	2–9
T. o. 'Golden Globe'	Golden Globe cedar	2–9
T. o. 'Holmstrup'	Holmstrup cedar	2–9
T. o. 'Holmstrup Yellow'	Yellow Holmstrup cedar	2–9
T. o. 'Pendula'	Weeping cedar	2–9
T. o. 'Repens'	Creeping cedar	2–9
T. o. 'Rheingold'	Rheingold cedar	5–9
T. o. 'Spiralis'	Spiral cedar	2–9
T. o. 'Techny'	Techny cedar	2–9
T. o. 'Tiny Tim'	Tiny Tim cedar	2–9
T. o. 'Vervaeneana'	Vervaeneana cedar	2–9

The author's dog, Finnegans Wake, sits in an ancient stool of *Thuja occidentalis*, eastern white cedar. The stool has made an elevated mound of many boles.

planted alone, they have wonderfully elegant vertical lines in the garden. Grouped together as a wall for the partition of garden spaces, they are unbeatable for their clean vertical planes.

There is a limited number of hardy weeping trees available to the gardening zones of 2–5. *T. o.* 'Pendula', or the weeping American arborvitae, is such a tree. It was developed about 150 years ago, and it is not seen much on this continent. Here in the garden I have a golden version of a chance seedling of a *T. o.* 'Pendula', which holds a very bright winter color. I have also developed a delightful vase-shaped dwarf *T. o.* 'Repens' that widens in circumference every year.

For more visual interest as a lawn specimen tree, *T. o.* 'Spiralis' is a dark green, pyramidal cedar with a spiral conformation of branchlets. There is also a cultivar with a distinct blue color, *T. o.* 'Techny' and a white variegated cultivar called *T. o.* 'Beaufort'.

Light green and yellow foliage can be found in combination in the cultivar *T. o.* 'Vervaeneana'. For a balcony or for a formal potted arrangement on a stone terrace the slow-growing *T. o.* 'Tiny Tim', a round green ball, becomes indispensable.

Two *Tilia neglecta* twinned for luck at the garden gate in spring

Tilia

BASSWOOD

Tiliaceae Zones 3–10

THE GLOBAL GARDEN

A tree that is called basswood by the North Americans is lime to the English, is *Linde* to the Germans, and is sacred to the Japanese, who planted *Tilia miqueliana,* the Chinese linden, around their temples. The linden tree is loved all over the world. Its sweet summer fragrance recalls memories of childhood play and endless sunshine that distills a past into a present sigh. *Unter den Linden* made Berlin famous not just for strolling relaxation but because most of the sixteenth-century manors had twin *Linden* at the front door for good luck. This practice crept over into the New World with the pioneers who erected farmhouses and planted a linden on either side of the front door step. This was done as a salutation to memory and hope for a good future in a new land.

While the little-leaf linden, *Tilia cordata,* has been part of the ancient wildwood of southern England, its big-leaf cousins have held the fort across the Atlantic. These are *Tilia americana, T. caroliniana, T. heterophylla,* and *T. neglecta,* the principal one being *T. americana,* or the American lime. These lindens are part of the ancient wildwood of North America. Because these species can also stool, regenerate from a living root system, they can grow into a ripe old age despite the Land Acts and carnage of settler occupation.

Collectively, the aboriginal peoples of North America are responsible for the naming of the *Tilia americana.* The Chippewa called it *Wigub'imij* because it produced one of the most essential items in their everyday life, fiber. *Bast* is an old name for fiber, and basswood named the tree. A discovery had been made back in the mist of time in North America that the bark of *T. americana,* if soaked in water in the shade, could be easily pulled into two separate pieces, one that could be plied into twine and further twisted into any gauge of rope. It could be boiled to strengthen the fibers into superstrong ropes. The fibers could be woven on looms into mats and even clothing. This fiber could be cut into widths as needed and stored as coils for everyday use. The string from basswood was used for tying and harvesting wild rice, in the packaging of medicines and food, in domestic sewing, for bag making, in making fishing nets, in stitching dome-shaped wigwams, and in threads, twines, and ropes. It was woven on a vertical wooden loom into fabric, using dry bulrushes as the warp and into beautiful carpet throws for flooring. These, interestingly, are similar to the basswood floor mats of the old Celtic world called *Cáiteog* from which the modern Irish word for throw, *caith,* is derived. It would seem that the mothers of invention were busy on both sides of the Atlantic Ocean, for no one likes cold feet.

The lindens make up a genus of about forty-five species of deciduous trees that are widely distributed in the north temperate regions of the global garden. There is one dominant linden in Canada, *Tilia americana,* the American linden. There is *T. neglecta,* the neglected linden. In addition, there are two further main species in the United States, *T. caroliniana,* the Carolinian linden, and the smaller *T. heterophylla,* the white linden with tomentose leaves. This tree is very similar to the silver lime, *T. tomentosa,* of Europe. All lindens have heart-shaped leaves and flower in midsummer. The flowers are produced in great numbers in drooping cymes that can be clearly seen from under the trees. These bisexual flowers produce a wonderful sweet fragrance that lingers for over two weeks until the flowers are fertilized. Each ovary produces a rounded, nutlike fruit called a drupe in the cyme clusters, again in huge numbers in the fall.

The *Smilax herbacea* vine, carrion flower, associated with *Tilia americana,* basswood, in the wildwood. The maturing fruit has already shed its impossible malodor and looks innocently beautiful nearby.

Tilia americana, American basswood at its best

After leaf drop, many of these clusters hang on the tree into the winter months and feed the overwintering seed-eating birds and mammals.

ORGANIC CARE

For the home gardener any of the North American lindens are easy to grow. These trees do well on a rich, fertile soil that is well drained. But they also show ample fortitude in growth on poor barren soils. They seem to have a preference for neutral soils. Their choice pH range is 6.5–7.5. They seem to also like a dolomitic subsoil, and this is where they find their natural habitat. So excessively acidic soils should be avoided.

Lindens transplant very well, taking about a year to reestablish new roots. Then they grow very rapidly, showing a high degree of drought resistance.

Many of the linden species sucker quite readily. In the case of these trees a single bole or trunk is sometimes produced and in another site for the same species a multiple trunk system is produced over the years. This compound trunk, in itself, is very attractive, but some gardeners object to it. They did in Europe around the seventeenth century. They cloned the lindens with single trunks for the grand avenues and allées that can be still seen today. The gardener can do this also. Any linden that is not prone to suckering can have a cutting taken as near to the ground as possible from a very low branch. This is taken in March, trimmed back, potted up, and allowed to grow on during the summer months.

All lindens can be grown from seed. Depending on the growing conditions in which the seed has been formed, it will show variable dormancy from one to three or more years. For the gardener, the path of least resistance is usually the most successful. The seeds can be collected from the ground or from the bare tree in the early winter. They should be immediately planted in sand or a half sand–half loam mix, ½ inch (1.2 cm) deep, in pots. These are buried in sand in a nursery row. For the colder zones

3–5, these pots can be covered with Reemay or leaves during the winter months. The pots should be well labeled because the seeds may take the full three years to break dormancy. When they do germinate, they are easy to handle. The seedlings can be separated readily in the spring and transplant well.

Lindens can be planted closely together for a pleached wall. Pleaching is an old method of weaving the branches of one linden into the next, to form a pleasing, woven fretwork that makes a living wall. This is kept trimmed on a regular basis, giving rise to a very attractive living architecture in any garden.

The North American gardener can grow all the linden species including the European and Asiatic ones. These seem to respond exceedingly well to high solar conditions. However, the reverse is not true. The European gardeners who wish to grow the North American species can expect a poor performance, but global warming might be on their side.

In general, lindens are remarkably disease free. There are some species of linden that are prone to the linden aphid, *Myzocallus tiliae,* such as the *T. platyphyllos,* the broad-leafed lime and all of its wonderful cultivars. Then there are others like the *T. tomentosa,* the silver lime, that kills these aphids in their tracks. Particularly in a warm and humid summer, the linden aphid will suck a great deal of sugar sap from the linden leaves. If these insects are very numerous, they will produce a honeydew carpet underneath the tree. If the summer is dry, without rain to wash off this carpet, it can be a bit slippery for walking and can even coat cars. So the planting of *T. platyphyllos* and all of its cultivars, the hybrid *T. p.* × *T. cordata,* and even *T. americana,* the American lime or basswood, and *T. neglecta,* the neglected lime, should be set some distance away from footpaths, driveways, and buildings over which these majestic trees can stretch a sticky limb in time. The other linden species are smaller and are usually groomed of aphids by the bird population.

In North American gardens, the Japanese beetle, *Popillia japonica,* an escapee from Japan, is spreading northward with the warmer winters of recent decades. The young and emerging leaves of *T. miqueliana,* the Chinese linden, was one of their favorite foods, and so their switch to the North American native linden species can be expected. During springtime, feeding frenzies take place overnight that can do great damage to lindens if the beetle population is high as it quite often is around new suburban areas. Doom (a trademarked name), the milky spore disease, can be applied to reduce the beetle numbers for the following year and to maintain a natural control on them in the garden. Not more than four grubs per square foot is considered to be the acceptable limit.

Tilia or linden flowers act like little umbrellas to protect the huge nectar flow from rain or morning dew.

MEDICINE

The medicine of the linden is to be found in the flowers, in the flower-winged bracts of which there is one produced for every flowering cyme cluster, in the fresh leaves, which are edible, and also in the bark.

The linden has been a sacred tree of the Indo-Europeans, who placed this tree next to the oak, *Quercus,* in high regard. The linden species they had used for medicines since prehistory were

Tilia or linden. The cream-colored flowers attract attention to the species worldwide.

Tilia cordata, the small leaf lime, *T. platyphyllos,* the broad-leafed lime, and a naturally occurring hybrid between the two major species, *T. cordata* × *T. platyphyllos,* sometimes called the common lime. They used these species in the management of fevers, as an expectorant, for hypertension and indigestion, for urinary tract infections, and for nervous states of anxiety and migraines.

The aboriginal peoples of North America used *T. americana* and *T. neglecta* interchangeably for the management of identical health problems, but these North American species would seem to have a higher component of pharmacological constituents than do the European species because it appears the aboriginal peoples used the basswood medicine for more serious situations.

All parts of the basswood, *T. americana,* contain volatile oil comprised of farnesol, α- and β-farnesene, and taraxerol, among other compounds. It also contains flavonoid glycosides, manganese salts, saponins, and polyphenols.

The flowers of *T. cordata,* the small leaf lime, were combined with the flowers of *Crataegus monogyna,* the common hawthorn, and those of *Viscum album,* mistletoe, in a very ancient European treatment for hypertension.

The Mohawk peoples treated internal hemorrhage with basswood, *T. americana;* the American elm, *Ulmus americana;* mountain maple, *Acer spicatum;* the speckled alder, *Alnus rugosa;* red baneberry, *Actaea rubra;* and goldenrod, the *Solidago* species. Both the inner and the outer bark of basswood were taken from the eastern side of the tree about 3.5 feet (1 m) from ground level and mixed with two handfuls of twigs about 5 inches (13 cm) long from American elm with one handful of twigs of speckled alder that had been smashed to release the outside bark. To this was added baneberry and goldenrod roots in 1 gallon (3.8 l) of water. This was boiled down to ½ gallon (1.9 l). Four ounces (113 ml) of this were taken every two hours until improvement was felt and then continued every three hours.

The Mohawk also used *T. americana,* the basswood, for mending broken bones. Leaves of *T. americana,* basswood, along with the leaves of *Eupatorium perfoliatum,* boneset; *Pinus strobus,* the white pine; and *Malva neglecta,* cheeses or common mallow; were added to fill a gallon. These were pounded and cold water added to make a poultice. This was placed on cotton cloth and cold wrapped around the break and was renewed every 24 hours for three weeks. Splints of white pine were bound over the boneset poultice.

A basswood tea, made by adding one teaspoon of fresh or dried leaves of *T. americana,* basswood, into 1 pint (.5 l) of boiling water, was used as a general panacea for soreness, for severe injuries, and as a general tonic. This was used by the Seneca peoples and is a common herbal tea used throughout North America.

The Iroquois aboriginal peoples treated serious burns and scalds with a poultice made by steaming equal volumes of leaves of basswood, *T. americana,* and *Fagus grandiflora,* American beech. These were mashed and applied to the burned surface.

They also used the basswood, *T. americana,* as a panacea for general body pain. The young growing branches of *T. americana* were picked in May or June. These were added to a small bunch of roots of *Epilobium angustifolium* of the evening primrose family, Onagraceae. This was steeped for 30 minutes in a gallon (3.8 l) of fresh well water. One cup (250 ml) of this was taken frequently during the day.

The Iroquois also had two cures, one for urine retention and its opposite, incontinence. *T. americana,* basswood, bark was used for urine retention. The bark was stripped upward. A bundle was made and boiled in well water. This was used for drinking. For incontinence, young shoots of *T. americana,* basswood, were steeped in well water. This water was used for drinking.

ECOFUNCTION

All of the basswood trees in North America are called a common name that seems to be used entirely in the rural areas. The *T. americana, T. neglecta, T. caroliniana,* and *T. heterophylla,* are all also called bee-trees.

They produce a huge crop of nectar. This nectar is second

The dainty woodland sunflower, *Helianthus strumosus,* usually found somewhere near *Tilia americana,* basswood, in a mixed forest

only to clover in volume for the honeybee. But what is more important, this nectar flow comes from all of the native species of basswoods at a vital time in the life of a hive. The queen bee has succeeded in laying and rearing her thousands of female worker bees to work the 10 square miles (26 km^2) of land around the hive. This is a quiet time for nectar from other plants but a time of high requirements for the hive. During the first two weeks of July, as the rest of the flowering world begins to dry out, the basswoods come into bloom. These trees produce an enormous flow of nectar in the evening hours and during the early hours of the morning until noon.

The design of each flower is such that the nectar produced on the inside of the sepal cannot be lost by rain or dew. This is because these flowers hang upside down and act like little umbrellas. The nectar, too, hangs in solution upside down. But the sepal produces a few fine hairs that, together with the physics of surface tension, are just enough force to hold the liquid in place without splashing down onto the grass underneath the tree. The form of the flower is such that it can easily be worked by the bumblebee, the honeybee, or any large or small flying insect who wants to drink. On a warm, sunny July morning,

Lilium regale 'Album', white regal lilies. They will amplify the fragrance of *Tilia americana,* basswood, flowering.

standing under a basswood tree is quite an experience. It is like sitting in the center of a honeybee hive.

Few people realize the importance of the honeybee to the agricultural economy of North America. The honeybee is an import but so, too, are the majority of the food crops on the continent. This insect is needed to cross-pollinate these crops, and of course, no crops means no food, means famine, all on the back of a honeybee.

Life's travails continue for the honeybee. It is a common occurrence in the heat of a summer's day that the hive needs to move on for one reason or another. This is when a decision is made to swarm. Quite often the swarm will head for the local ancient basswoods. There they dig out the soft dead wood of the trunk to make a cavity. They set up house, with the honey stores overhead. They stay there for a few years, until a family of raccoons or even a bear literally comes knocking. The cavity is taken over by either porcupines, flying squirrels, overwintering squirrels, or one of the larger birds for nesting. The basswood becomes a succession of habitats for many creatures.

In some years, *T. heterophylla,* the white lime, and in particular, *T. tomentosa,* the silver lime, produce a hypnotic biochemical for honeybees. These trees both produce stellar glandular hairs on the undersides of their leaves that give them both a silvery white flash when there is a wind. The foraging bees go into a hypnotic trance and fall under the trees in large numbers. These trees appear to produce an as yet uncharacterized sedative chemical. The bees die and then act as a nitrogen fertilizer for the tree. The honeybee, if removed from this situation, will recover. Under these conditions both *T. heterophylla* and *T. tomentosa* are both functioning as carnivorous trees. This phenomenon, to say the least, is unusual in the plant kingdom.

Because basswoods produce a large quantity of nectar and honey dew, the trees are also very attractive to aphids, except, again, for *T. tomentosa* and *T. heterophylla,* which both produce toxins for aphids.

These insects protect their own well-being in a unique way. When in danger, they produce a series of isomers of a very complex architecture called trans-trans and cis-trans. They twist an alarm biochemical called farnesol into these volatile forms. The aphids then cast this biochemical onto the waves of flower fragrance around the basswood, like a cork on the ocean surface. The farnesol alarm signal travels on waves of (*E,E*)- and (*Z,E*)-, two isomeric forms of α-farnesene out into the forest airways, warning other insects of imminent danger.

There are a number of higher-order mushrooms associated only with the *Tilia* species. Some of these are *Crepidotus tiliophilus* and *Hebeloma velatum.* This association is not understood.

BIOPLAN

All of the basswood species are becoming rare in North America. Trees that are over a hundred years old with diameters of 6–7 feet (2 m) are decided oddities. Since the settlement of North America the basswood has been in decline mostly through cutting as weed trees on the basis that the heat produced by the wood is less than half that of oak or hickory. But in a north temperate forest, the basswoods are feeding trees for the flying insect world. Insects are beneficial. They represent predation and prey within the multitude of other arboreal species that depend on the forest basswoods for health. Many birds feed entirely on insects, so they, too, follow in the footpath of the basswood. Basswoods feed the forest; they must find their place in regeneration programs alongside the stronger trees if there is to be a balance of nature.

Because the basswoods, in particular *T. americana,* the American linden, and *T. neglecta,* the neglected linden, can be coppiced easily, the living stools of some of these species must be extremely ancient, if they were to be identified. Some of the larger stools might well be one thousand years old in North America.

The basswoods could be coppiced as a source of high-tensile bast for nonallergenic carpeting, for roping and ties in the horticultural world, and for renewable organic packaging as a substitute for paper products.

The majesty of the basswood, in particular *T. americana* and *T. neglecta,* could be used to great advantage in parks, as shade trees for wide roadways, in housing estates, and as city walking allées, much as their cousins were once used over the past centuries in Europe.

A bioplan for beekeepers would supply a sequence of honey flow for most of the summer months from June into August. *T. platyphyllos,* the broad-leafed lime, flowers in the middle of June. *T. cordata* × *T. platyphyllos,* the common lime, and *T. caroliniana,* the Carolina basswood, flower at the end of June. *T. maximowicziana,* the Japanese lime; *T. americana,* the American basswood; and *T. heterophylla,* the white basswood; all flower in July. *T.* × *orbicularis,* the hybrid lime, and *T. miqueliana,* the Chinese linden, both flower in August. This kind of extended bee pasture would only work in North America and would be a boon to beekeepers.

Basswoods should be part of the farm bioplan. These trees should be part of a hedgerow system or could line the roads around the farm buildings themselves. In any case, these trees will attract and feed both wild and cultivated honeybees, which will result in an increased yield of crops.

Basswood lumber is a high-quality wood for carving. The larger basswood blocks for sculpting are becoming rare in the art world. A set-aside scheme for farmers encouraging the planting of basswood as a future cash crop, especially since they can be coppiced, would benefit habitat regeneration and provide long-term income.

DESIGN

In a north temperate garden, especially in zones 2–5, it is a challenge to design a fragrance garden. This is because of the relatively few species available to the gardener. Most of the basswoods are a good answer. They step up the airways with fragrance during the summer months in an invisible way. The variety of trees this group represents is large, going from the smaller, almost dwarf, tidy trees of *Tilia cordata* 'Greenspire' and *T. c.* 'Rancho' to the majestic 100-foot (30 m) flowing form of the native *T. americana,* the American basswood. In between there are the *T. caroliniana,* a more chunky tree, and the flashing white fire of *T. heterophylla,* a fashion fit for any white garden. At the top of the list is probably the most beautiful weeping tree in North America, *T. a.* 'Pendula'. This tree has a rounded domed crown with downward-sweeping branches and is sweetly scented. This species is of uncertain origin, appearing on the American market in 1840. The underleaf surfaces have a similarity to *T. tomentosa.* They are like white velvet, ruffling with the slightest breeze. There is also a rare columnar form of our native *T. americana.* It is called *T. a.* 'Fastigiata', the fastigiate basswood, which was developed in New York in 1927. It has a narrow conical form with wonderful abruptly ascending branches.

All of the Tiliaceae or basswood species are ideal species trees

A number of higher order mushrooms are associated with *Tilia,* basswood, such as the coral fungus, *Clavaria aurea.*

for homes, schools, and nurseries, in fact anywhere there are children. This is because both flowers and leaves are nontoxic to man. When the tree is in the act of feeding insects in midsummer, these insects feed from the top downward and are out of reach of a child. In addition, a feeding honeybee is not a stinging honeybee; they are programmed by fragrance for food and not for defense.

If the Japanese, Chinese, and Manchurian lindens, *T. maximowicziana, T. miqueliana,* and *T. mandshurica* are being used as part of a garden design, they should be placed in a sheltered spot that has some spring shade to protect the foliage from late frosts. The shade will delay the buds from elongating prematurely. These species trees will flower from July into August.

The fragrance lilies, *Lilium regale* 'Album', especially when planted in a mass, will both amplify the fragrance of the basswood and will very subtly underline the reflexed form of both the large white lily and the basswood cyme. Both listen to the same call for fragrance, and both flower at the same time.

Another native can be mixed with any of the basswood species for background fragrance that acts to enhance the airways around a garden. It is the lowly fox grape, *Vitis labrusca,* for which the entire continent was once name Vineland. This grape produces hauntingly fragrant, greenish, small flowers in June and into July.

Tilia Species and Cultivars of Merit

SCIENTIFIC NAME	COMMON NAME	ZONES
Tilia americana	American basswood	3–9
T. a. 'Dentata'	Coarse-leafed basswood	3–9
T. a. 'Fastigiata'	Fastigiate basswood	3–9
T. a. 'Pendula'	Weeping basswood	3–9
T. caroliniana	Carolinian basswood	4–9
T. cordata	Small-leafed lime	4–9
T. c. 'Greenspire'	Greenspire lime	4–9
T. c. 'Rancho'	Rancho lime	4–9
T. heterophylla	White basswood	4–9
T. mandshurica	Manchurian lime	3–9
T. maximowicziana	Japanese lime	4–9
T. miqueliana	Chinese linden	5–9
T. neglecta	Neglected basswood	3–9
T. platyphyllos	Broad-leafed lime	4–9
T. p. × *T. c.* (syn. *T.* × *europaea*)	Common lime	4–9

The somewhat elongated crown of a mature *Tilia americana*, American basswood

Tsuga canadensis, eastern hemlock. It thrives in a mixed hardwood forest because of the cathedral-like canopy that produces just the right amount of gloom for optimal hemlock growth.

Tsuga

HEMLOCK

Pinaceae

Zones 3–10

THE GLOBAL GARDEN

Like Snow White, the hemlock is the fairest of them all. In the forest the hemlock is the most elegant of the evergreens. It has a naturally graceful, weeping form, flowing from the forest floor in an evergreen fountain and back into it again in a river of shady growth. Throughout the long history of the global garden the hemlock is a true survivor. After the carnage of the last ice age it has managed somehow to keep the faith in eastern Asia and North America. The hemlock was wiped out in Europe. In Poland, France, and England traces of hemlock are only to be found as fossil pollen. This pollen is buried deeply and is the only headstone of an ancient habitat.

The hemlock is reduced to just 10 species. At the top of the list is the familiar Eastern hemlock, *Tsuga canadensis,* which crowds around the Great Lakes and, both great and small, grows toward the eastern seaboard ending at the top of King's County, Prince Edward Island. Another hemlock was discovered about one hundred years ago growing on the Blue Ridge Mountains, the Carolina hemlock, *T. caroliniana.* The other two American species are giants of the west coast rain forest. The west coast hemlock, *T. heterophylla,* grows to 250 feet (76 m) in 500 years. The other hemlock clads the western mountains up to a dizzy 11,000 feet (3,350 m), tracking from Alaska down into California. It is the mountain hemlock, *T. mertensiana.* Across the Pacific are the six Asiatic cousins, who are mostly shrubby in growth, the most important being the Himalayan hemlock, *T. dumosa,* which can grow to 100 feet (30 m) in sheltered mountain valleys. A rare species is the Taiwan hemlock, *T. formosana,* with short leaves and tiny cones.

From an evolutionary point of view the hemlock is a very ancient tree. What was lost from the genetic storehouse of this tree by extinction we will never know. It is an important tree of the virgin forests with an unusual mandate, that is, to have regenerative synchronicity with maturing trees. Just for this reason alone, the hemlock should not be reduced to terms of crates holding produce for the local supermarket.

ORGANIC CARE

The hemlock is a tree that adds prestige to any garden. It is a tree for a shade garden. It will not grow well in an open area in full sunlight because it can get both summer and winter sunburn. It is perfect for a garden with a north-northeasterly exposure, one that is difficult to design for at the best of times. Sulfur dioxide will damage the tree, so it should be located away from highways or roads with dense vehicular traffic. This evergreen is one of the few trees that can be planted as a young companion tree with mature spruce, maple, and beech.

It is important to know that hemlock trees do not travel well. If the hemlock is to be planted as a seedling tree, it must be obtained from a local nursery that, in turn, has grown it from local seed stock and not obtained it, as seems to be the usual set of circumstances in Canada, from a southern source. If this happens, the southern hemlock has too long a growing period for its northern site and will be subject to frost injury. However, a northern tree set into a southern garden will have a very short growing year that will not harm the tree.

Hemlock can be grown from seed. The seed itself is triangular in shape and comes with one wing. The seed has its own resin pockets, which are a tiny arsenal of chemicals to keep disease at bay from the young growing seedling. Hemlocks bear seed at around thirty years of age. For the collection of seed, a tree of 30 feet (9 m) should be selected. This tree will have an ample selection of tiny cones within easy reach. The cones should be collected either around or after the middle of September, when they begin to mature. Depending on the year, it is important to be able to read the cone color for collection. A ripe cone of *Tsuga canadensis,* the eastern hemlock is purple-brown. *T. caroliniana,* the Carolina hemlock, is biscuit brown. *T. heterophylla,* the west coast hemlock, is brown with red-brown tips. *T. mertensiana,* the mountain hemlock, is dun-brown.

A good selection of cones is collected. These will be male and female cones. The female cones will contain ripening seed. The cones are placed on dry newspapers in a warm, well-ventilated room for a minimum of two days. The seeds will slip out of the cones with a light tap. The seeds can be stored at 41°F (5°C) in paper envelopes packed into glass containers in a refrigerator.

Before planting in the spring, the seeds are left in full sunlight for 12 hours. They are then surface planted in pots or into a nursery row. The seeds are covered with burlap that is kept damp until the seeds germinate. When the radicles or young roots have penetrated the soil, the young seedlings are mulched with 1 inch (2 cm) of peat moss that is also kept damp. The first leaves, or cotyledons, are visible in four days and produce three to six leaves after seven days. The nursery row or pots should be draped with two layers of Reemay during July and August. One layer of Reemay can be left on the seedlings during their first winter for both shade and to prevent winter desiccation. Seedlings are usually transplanted the second or third spring.

Hemlocks like a rich, well-drained forest soil. The soil should be on the acid side, pH 5.0–6.5. Peat moss can be added to the planting hole to acidify the soil. The eastern hemlock, *T. canadensis,* is fairly lime tolerant, as is the Himalayan hemlock, *T. dumosa.*

Although hemlocks appear to be sensitive to ozone, they are the most disease-free of all of the evergreens. Rhododendrons and hydrangeas should not be planted near hemlocks because they cohost a blister rust caused by *Pucciniastrum vaccinii* and *P. hydrangeae.* There are six or so cankers of the hemlock, which can be controlled by judicious pruning. In some areas of eastern Canada, there is a heartwood rot caused by two fungi. This fungal infection cannot be cured. This disease causes the death of the tissues underneath the bark, which kills the tree. Hemlocks tolerate neither heat nor extreme drought. Hence a shaded or northerly positioning of these trees and their cultivars in a garden will keep these species healthy.

MEDICINE

In many ways it is unfortunate that the hemlock became extinct in Europe. As a member of the pine family it undoubtedly has medicines that need further investigation. It has long been used in North America by the aboriginal peoples as an emergency food source, in the treatment of fevers, colds, sore throats, general infections, and rheumatism, as haemolytic agents, and in the management of various venereal diseases. These are very similar to the varied medicinal uses for the pine in the many ethnic pharmacopoeias of Europe over the past centuries.

Around Prince Edward Island, home to the Mi'kmaq, these aboriginal people made considerable use of their eastern hemlock, *Tsuga canadensis.* They made a tea of the living inner bark of the tree. It was used for the treatment of pain and colds. Further south, the Menominees had similar medicines.

The leaves were used by the Potawatomis peoples to make a tea that induced copious and excessive sweating, thus breaking a cold. This is a very ancient technique used to treat colds. It is also effective and rapid, taking about twenty-four hours.

The Ojibwa peoples used the bark of *T. canadensis,* which is very high in tannic acid, to heal open wounds and to stop the bleeding.

The Seneca peoples used the leaves and bark of *T. canadensis* in the treatment of rheumatism. A handful of leaves and bark was put into 6 quarts (5.7 l) of well water. It was boiled until steaming. A hot, heavy object was placed into the water keeping it steaming. A tent, made with a blanket, was placed over the pot. The patient inhaled the steam under the tent for as long as the steam lasted. The patient's feet were then placed into the lukewarm water until it cooled. This was repeated three times a week.

Hemlock was used by both the Seneca and Cayuga peoples for the treatment of cholera. The fronds of the polypodum fern called the rock polypody, *Polypodium virginianum,* which is found on rock faces in mature deciduous forests, was combined with the leaves of eastern hemlock, *T. canadensis,* and chips of bark of swamp white oak, *Quercus bicolor.* The finer details of this treatment have been lost.

The Mohawk aboriginal peoples used eastern hemlock, *T. canadensis,* for the treatment of the early stages of tuberculosis. Both roots and young fronds of the sensitive fern, *Onoclea sensibilis,* were taken together with the bark of witch hazel, *Hamamelis virginiana.* This bark was peeled upward. Chips of the outer bark of eastern hemlock, *T. canadensis,* were added to the roots of bracken fern, *Pteridium aquilinum,* and mixed with the young shoot tips of balsam fir, *Abies balsamea,* and the young shoot growth of the rare Canada yew or ground hemlock, *Taxus canadensis.* This was put into 8 quarts (7.6 l) of water, which was reduced to half and strained. The decoction was taken as needed.

From a nutritive point of view, the young growing leaves that are still somewhat blanched make an excellent tea that is high in vitamin C. The inner living bark can be dried and pounded into flour. This can be added to other flours or used alone.

ECOFUNCTION

Within the grand scheme of the forest the hemlock has a considerable ecofunction. In an eastern Ontario wildwood that has an age of one hundred years or more, the hemlock is a specimen tree of all ages dotting the landscape around the mature, large trunks

Stealing the show. A young eastern hemlock, *Tsuga canadensis,* gets a toehold in a young, mixed hardwood forest.

Tremella reticolata, white coral jelly. Of the many fungi associated with *Tsuga canadensis,* eastern hemlock, this, with its human-brain-like surface, is most intimidating.

of older trees such as the maples, beech, and hornbeam. Apart from the cedar, it is, in the wintertime, the only evergreen sweeping the ground in very close contact with these trees. It is a tree of patience. It bides its time until the mature hardwood trees die. Then it rockets up in growth, forming a damp canopy as a nurse crop for the propagation of maple, beech, and hornbeam in a long, repeating cycle of growth.

Twinned with this is another phenomenon of the woods. In the winter months when the temperatures drop, many animals enter into communal living. One of these is the porcupine, *Erethizon dorsatum.* These creatures hole up in groups of five to ten in the cavities of mature maples. In the early days of spring as the sun gets higher on the horizon and their dens warm up, the porcupines will head for food. The nearest hemlock will get a severe trimming of its foliage. This will, in turn, damage the tree, loosening the bark, making this particular hemlock prone to one of twenty varieties of bark borer. Later on in the spring a migratory bird drops by its favorite haunt. This is a yellow-bellied sapsucker, *Sphyrapicus varius,* a member of the woodpecker family who prefers the hemlock and the birch over other trees. The sapsucker gets down to work. He or she punctures the bark right into the living cambium, where a literal waterfall of sugary food is released. The sapsucker feeds on this and lets it flow for a while. This river of food attracts a host of hungry insects, an additional dinner bonus for the sapsuckers, who also enjoys the odd bark borer. When the sapsucker gets very enthusiastic, he or she girdles the tree. This begins the tree's demise, which in turn attracts the interest of the pileated woodpecker, *Dryocopus pileatus,* who sets up his rectangular, holed residence for nesting and ringing his song throughout the forest. In time, punctured by the cavities of many homes, the hemlock falls, to die and decay in an extraordinarily rapid time, creating a soil ideal for its own seedlings. This nursery soil is spongy and damp. This, together with filtered shade, is the perfect design for young hemlock growth. The cycle begins again.

During the hungry winter months the hemlock is a feeding tree for deer, rabbits, snowshoe hares, partridge, and grouse.

The western species of hemlock are very important for erosion control on contoured land. This is important for clear water that, in turn, has a high oxygen content. These pools and mountain streams are the birthing ground for fish such as salmon, who switch hormones and substitute freshwater for the saltwater tides in the long cycles of their wandering lives.

The design of the leaf of hemlock is self-protecting with respect to global warming. Unusual among the conifers, the hemlock has a leaf petiole. This places the hemlock at an advantage over other evergreens who are sessile and do not have a petiole, in that the leaf can reorient itself for more or less sun, to meet the needs of the tree. The upper surface of the leaf has a thick, cuticular layer that is xerophytic and will not allow a high degree of water loss. This is enhanced by its parabolic shape in cross-section. On the undersurface of the leaf lies the breathing apparatus of the tree called the stomata. These are in rows and are also in turn protected from excessive moisture loss by a fine coating of hairs. These are adaptive characteristics of more evolutionarily advanced conifers or evergreen trees.

All of the hemlock species are very ancient trees. They have tasted global warming before. The North American and Asiatic species have survived by modifying their own genetic code. The hemlocks, *Tsuga* species, now have a very specialized molecular clock regulating time and place. In other words, the hemlock seems to have a global positioning tool that is linked with time. The hemlock does this by means of enzymes that have just recently been discovered. They appear to be very elegant molecular tools for survival. These kinds of tools are like serotonin and melatonin in man, which are tied to migraine, Parkinson's disease, and schizophrenia.

The hemlock, as a member of the pine or Pinaceae family, plays host to the largest number of mushrooms of the north temperate forest. This may be indicative of the hemlock's ability to act as an underground cofeeding tree to the mature tree neighborhood through mycorrhizal food exchange.

BIOPLAN

The north temperate wildwoods show a different vision of the hemlock's place in the forest. Casually walking through a deciduous forest in winter, one is struck by the natural placement of the hemlocks. They are of all sizes and ages. They show a pattern of seed dispersion and growth that is interesting. The trees grow in a singular fashion, always well spaced from one another, almost in a grid fashion, and almost always in close contact to a mature deciduous tree trunk.

The hemlock is being ignored in modern reforestation. This may not be good for the future of our forests. The hemlock should be part of the bioplan for forest regeneration. The hemlock is a gregarious species. It needs the presence of other trees in close proximity in order to live and thrive.

There may also be another scientific factor in the hemlock's growth that is being completely missed. The hemlock is a shallow-rooted tree. Its companion trees always have deeper roots going into the subsoil strata. Because the hemlock is associated with such a great diversity of fungal species that are capable of mineral solubilization, it is possible that there is a mineral exchange going on between the hemlock and the neighboring trees and that some alleloprotection is being conferred on the deciduous trees also.

Equisetum scirpoides, dwarf scouring rush, associated with the moisture of *Tsuga canadensis,* eastern hemlock, making its statement continuing evolution. Because of logging, this plant of ancient heritage is rarely seen.

DESIGN

The deep, rich, forest green color of the hemlock, *Tsuga* species, does not show seasonal change, making these trees a foil for the winter landscape. As a specimen tree for the shade, there are few

Mertensia virginica, Virginia bluebells, an ideal flowering companion to any *Tsuga,* hemlock, species, large or small. This fragrant, hardy native disappears after flowering in the spring.

trees more delicately elegant for the garden. The eastern hemlock, *T. canadensis,* is perhaps the most beautiful, with the Carolina hemlock, *T. caroliniana,* a close second. Both of these trees are rarely seen in North American gardens but are commonly used in England and Europe. These trees are fairly drought resistant and frost hardy. The eastern hemlock will tolerate a neutral to alkaline soil, while the Carolina hemlock prefers a more acidic soil, as do all of the other hemlocks except for the Himalayan species.

In the garden, the eastern hemlock has a floating appearance. The tree seems to be in constant movement, starting with the weeping tip. In a more open situation the lower branches sweep the ground in a most attractive way. The bobbing of myriads of tiny cones on the foliage accentuates the flowing appearance.

The western hemlock, *T. heterophylla,* is used extensively as a single specimen tree in nonchalk soils. This tree is very fast growing with delightfully graceful branches. It has a particularly fine cultivar, *T. h.* 'Greenmantle', which originated in Windsor Great Park in England.

The eastern hemlock, *T. canadensis,* has a large number of cultivars that are enormously useful to the gardener. These are the mounding hemlocks that are spectacular for trough or rock gardens. Once a site with cool roots and some shade from the winter sun is chosen, then the cultivar will be happy for a long time.

Probably one of the most remarkable cultivars is *T. c.* 'Cole' (syn. *T. c.* 'Cole's Prostrate'). This plant produces extensive carpets of long, flattened branches. This is ideal for water garden design because the leaves will not fall and affect water quality.

Another wonderful cultivar is *T. c.* 'Pendula' (syn. *T. c.* 'Sargentii') or Sargent weeping hemlock. This is a weeping form with contorted branches that shows a high tendency to reproduce true to seed. It is one of the most striking specimen evergreens for a lawn or a rock garden.

For the smaller city gardens there are a number of dwarf cultivars. These include, *T. c.* 'Jeddeloh', which has branches arching out from a depressed center, *T. c.* 'Lutea' with golden foliage,

Tsuga Species and Cultivars of Merit

SCIENTIFIC NAME	COMMON NAME	ZONES
Tsuga canadensis	Eastern hemlock	4–9
T. c. 'Cole'	Cole's dwarf eastern hemlock	4–9
(syn. *T. c.* 'Cole's Prostrate')	(Cole's prostrate eastern hemlock)	
T. c. 'Dwarf White Tip'	White tip dwarf eastern hemlock	4–9
T. c. 'Jeddeloh'	Jeddeloh eastern hemlock	4–9
T. c. 'Lutea'	Golden dwarf eastern hemlock	4–9
T. c. 'Sargentii' (syn. *T. c.* 'Pendula')	Sargent weeping hemlock	4–9
T. c. 'Swain'	Swain eastern hemlock	4–9
T. caroliniana	Carolina hemlock	5–9
T. heterophylla	Western hemlock	5–9
T. h. 'Greenmantle'	Greenmantle western hemlock	5–9
T. mertensiana	Mountain hemlock	5–9

Leotia viscosa, winter slippery cap. Chasing spotlights in the gloom of a mature forest within the root system of *Tsuga canadensis,* eastern hemlock, in August, *Leotia viscosa* sings its song.

T. c. 'Dwarf White Tip', and a particularly hardy cultivar *T. c.* 'Swain'.

All of the hemlock, *Tsuga* species, combine well with large rock surfaces, water, and expanses of flat lawn. These species also combine with club mosses like *Lycopodium* and *Selaginella* with *Isoëtes* species, quillwort, placed in the cracks of rocks. Many varieties of *Equisetum* look good with hemlocks, particularly the horizontal kinds like *E. telmateia* and *E. sylvaticum.*

The golden glow of evening light, licking the face of a community of *Ulmus americana,* American elm

Ulmus

ELM

Ulmaceae

Zones 3–9

THE GLOBAL GARDEN

The landscape of the eastern portion of North America was made famous by the elm and was immortalized by the canvases of A. J. Casson and the Quebec painter Marc-Aurèle Fortin. In the canvases, the elm stands proud and free, dappling farmhouses, cities, and streets with snowflakes of colorless shadows.

The elm carries an important historical documentation on the civilization of man. It is in the form of a disease. Nowadays it is called Dutch elm disease. This disease is not new. Pollen analysis shows that the European Neolithic revolution of agriculture that took place around 4500 B.C. coincided with an increased death rate in the elm. The tree recovered to flourish. It took a death dive again with the Bronze Age around 2400 B.C. There was a recovery and then another dive with the Iron Age around 500 B.C. During the past 50 years leading up to this age of technology, there has been another decline in Europe with the elm, again, showing some resistance.

For the North American elms, their continental isolation kept them pristine until recent times. However, beetles, unknown to their human transporters, were imported, infesting the planks of a coffin in the 1920s. The disease has spread throughout the continent, traveling in infective waves into the forests when the conditions were right for its spread. Dutch elm disease has been particularly virulent in North America because it is a "new" disease. However, the trees are reverting to an age-old defense mechanism, that of resistant, underground root suckering. The roots may lie dormant for up to twenty years after the death of the parent tree. These suckers sprout into rapid growth, the great majority of which shows a satisfying degree of resistance to Dutch elm disease. What is new is old for the cycles of life of the elm. This is why there is resistance in European species and before that in Asiatic elms.

To the rural mind the elm is a "widow maker." This is the farmer's nickname for a dead or dying elm whose branches crash with great abandon. The wood is needed for firewood, and a dance of death is performed by the logger in trying to bring it down. Anyone who has helped pile a winter's wood supply will be familiar with the tracks of the beetles responsible for the transmission of the Dutch elm disease. The elm bark separates readily from the dead wood. This exposes, on the inside of the brown dusty bark, a series of quite beautiful, lacy tunnels. These tunnels are the remains of the homes of the bark beetles, *Scolytus scolytus* and *Scolytus multistriatus*. The female lays her eggs at the dead end of each tunnel. A grub hatches and tunnels sideways, causing a lot of debris and soft compost. This is the ideal dark garden for the spores of a fungus, *Ceratocystis ulmi*. The spores grow and reproduce. As the grubs turn into beetles, they walk over this fungal carpet, scuffing off the highly pathogenic fungal spores on their feet. The beetles keep walking toward light. When they reach the end of the tunnel they fly to another elm and start the cycle of disease all over again. However, the story can sometimes get worse. The fungus itself may carry a disease called *Morus ulmi*, a mycoplasma, which collapses the phloem tissue of the elm. This is lethal to the tree. Death may be very rapid, sometimes occurring within the week, beginning with the wilting and browning of the limbs and progressing throughout the tree.

There are about thirty-six elm species in the north temperate regions of the global garden. In almost every country, a native elm is to be found close at hand. Some of these include, in Russia, the Russian elm, *Ulmus laevis*, and the Siberian elm, *U.*

Delectable and delicious, *Morchella angusticeps*, the narrow-capped morel

Curiosity

Joe Smoke was the aboriginal name I heard again and again over the years. This man came into the Wolford, Ontario, area in the 1920s looking for slippery elm, *U. rubra*. He came twice a year, once in the spring and again in the early fall.

This kindly man, Joe, allowed my neighbor, then a child of eight years or so, to watch him gather in the woods. Joe Smoke collected elm bark in the spring for medicine. He also collected morel mushrooms, *Morchella* species, and in the fall ginseng, *Panax quinquefolius*. His peoples had been doing this for thousands of years.

Joe Smoke was the last of his line, but he allowed the curiosity of a little eight-year-old boy to be kindled, to watch and to remember . . .

pumila; in Japan, the Japanese elm, *U. japonica;* and in the British Isles, the Scotch elm, *U. glabra,* also known as the Wych elm. In the southern United States there is a September flowering elm, *U. serotina*. There is also a little, flesh-leafed elm called the cedar elm, *U. crassifolia,* going as far south as northern Mexico. All over the eastern portion of the continent the enormous American elm, *U. americana,* can be seen. There is also a rock elm, *U. thomasii,* and the slippery elm, *U. rubra,* both of which are becoming rare.

Elm wood has been used for rural furniture much as deal wood was used in the kitchens of the past in Ireland, England, and Wales. The furniture from rock elm, *U. thomasii,* is of very high quality. Few North American farmers worth their salt are without their own personal hoard of elm planks for emergency situations. This is because, despite a smell of urine, the planks are strong and hard and will hold a nail exceedingly well if green and in a pinch will burn with a greater heat output than coal, if they get split pulling a tractor out of snow or mud.

It is the North American aboriginal peoples who have made the elm famous on this continent. In particular, within a forest they knew where every slippery elm, *U. rubra,* existed. They used the bark of this tree for medicines. It was extensively used as a source of fiber, which was separated by impact blows to the bark. The tree fibers separated longitudinally and were woven into fine and coarse threads, into string and further into rope. All of these items made from the tree's fiber or bast were essential goods in everyday life. Having harvested the bark, which sometimes killed the tree, the aboriginal peoples then went on mushroom hunts after spells of humid weather from the late spring into the summer for the most delectable food item on the North American continent, the morel. The true morel, *Morchella esculenta;* the delicious morel, *M. deliciosa;* the thick-stemmed morel, *M. crassipes;* the conic morel, *M. conica;* and the rare gem, the two-spored morel, *M. bispora;* are all found associated with *U. americana,* the American elm. Some of these they dried for winter, and others they used in their cuisine.

The Iroquois aboriginal peoples also used the bark of the slippery elm, *U. rubra,* for their canoes. In the midspring, when the sap was rising in the elm, they slipped the entire bark from the trunk of a larger elm. This they turned inside out, shaped, and, using the elm bark mucilage as caulking to keep the vessel watertight, fashioned into canoes. These elm canoes worked well in daily life but when used as war machines were left sitting in the water while their enemies, the Hurons, developed considerably greater speed with their state-of-the-art birch bark canoes. War and peace was thus decided with the elm.

ORGANIC CARE

The elm is not being actively replanted in North America. Unfortunately over the past 30 years, municipal authorities had the opinion that any elm should be a dead elm. So the rising resistance in the elms was destroyed. However, recently, throughout the continent all the elm species are making a comeback.

For the gardener, resistance can be seen in an elm grove, if when most of the trees have died due to Dutch elm disease, a few, old, faithful trees remain. These resistant species should be sought out and propagated.

All of the elm species can be grown from seeds that are also called samaras. Apart from *U. serotina,* the September-flowering elm, the elm flowers in a rush in early spring, sometimes even before the leaves emerge. The circular fruit complete with ripening seed is produced during the early summer months. In many cases these can be picked from the lower branches. A particularly sharp eye has to be kept on the rock elm, *U. thomasii,* because the ripe seeds are especially valued by wildlife. As soon as a few elm fruits begin to fall, these seeds are ready to be harvested in the midsummer months even though they are green on the branch.

The freshly collected seeds are strewn out on newspapers in a dry room for two to three days. The seeds turn a mud brown. They can be immediately sewn in sandy loam covered by ¼ inch (.6 cm) of soil, about 2 inches (5 cm) apart.

Elm seeds are all circular packages with varying density and hairiness. The embryonic seed inside this package is sometimes quite fragile, as is the case of the American elm, *U. americana.* So when the seeds are being stored, the entire samara is left intact. They are stored in sealed, plastic bags to both retain sufficient moisture for germination and to keep the samaras themselves in a flat position. Care should be taken to keep all elm seeds from becoming too dry or they will lose their ability to germinate.

All elm seeds show some dormancy. The more northern species, such as *U. rubra,* the slippery elm; *U. thomasii,* rock elm; *U. americana,* American elm; and *U. villosa,* the Himalayan elm; require

The landscape of North America, made famous by the elm, in particular *Ulmus americana,* American elm

Ulmus americana, American elm, a sentient being that stands as witness to the ephemera of a garden. This tree is one of many showing resistance to Dutch elm disease.

8–12 weeks of stratification at 41°F (5°C) in the crisper of the refrigerator. If local seeds are being collected in zones 2–3, these seeds require the maximum length of time to break dormancy.

Each seed complete with its lopsided wing can be potted up after stratification. The seeds are covered with ¼ inch (.6 cm) of soil. Usually they germinate in about ten days. These young seedlings look somewhat like young apple seedlings with twin cotyledons at the center of which two serrated leaves emerge. The seedling elms are easily transplanted in their first or second year. Once they have become established, an elm may grow up to 12 feet (4 m) in one growing season.

Occasionally, in late spring, especially before a hot summer, the young fresh leaves of elm fall prey to the feeding frenzy of the mourning cloak butterfly, *Nymphalis antiopa.* The spring caterpillars can be identified by having a row of red dots on their back. They line up in army fashion to feed in parallel rows. The butterflies that emerge from the chrysalides should be welcomed into any garden.

Apart from Dutch elm disease, many elms are also commonly affected by another disease called wetwood. This disease seems to come and go in these trees in a cycle that is not understood. Wetwood becomes evident when the sap begins to flow from spring to fall. In one of the lower crotches, a dark, watery exudate of high pH is seen to seep into the bark and down the trunk to the ground. Sometimes it has a green, thick character caused by populations of the unicellular green algae called *Chlorella.* Wetwood is caused by an aerobic bacterium that, in turn, makes the wood resistant to fungal infection. This disease is chronic and its parameters are not understood. It has to be lived with both by the gardener and forester.

MEDICINE

The slippery elm, *U. rubra,* has been a very old and popular medicinal tree of the aboriginal peoples of North America. The American elm, *U. americana,* has also been used. Its Asian counterpart, *U. campestris,* also known as *U. minor,* the smooth-leafed elm, is used in Chinese medicine and was introduced into England during the turbulent Bronze Age.

The bark of the elm is used for medicine. The bark is harvested after the sap has begun to flow. On the North American continent this time varies from late January to February in the south and can be delayed into the beginning of May in some years in the north. The area of the bark that contains the medicine is the inner bark or the endodermis of the tree. This endodermis has a mucilage flow that is fragrant. It has the odor of fenugreek, *Trigonella foenumgraecum.* This endodermis also extends down into the root tissue, and it, too, is occasionally used. The mucilage flows until midsummer and is somewhat similar to the pectin-rich gel of the flesh of a grape.

The mucilage, in its gummy matrix, carries a chemical load of sesquiterpenes, alkaloids, steroidal sapogenins, a variety of sugars, tannins, calcium oxalate, vitamins A and C, cobalt, selenium, silicon, tin, and zinc. The mucilage itself is composed of a large polymeric formation of α-D-3-methyl-galactose sugars, which are also known as brain sugars. These sugars probably occur as polygalacturonic acid among other polymeric forms of glucose, fructose, hexose, *l*-rhamnose, and pentose in various amounts.

The bark of slippery elm, *U. rubra,* was commonly used to induce abortions by the aboriginal peoples. It is now used medically, in its powdered form, to induce abortion.

The Seneca aboriginal peoples used slippery elm, *U. rubra,* in the treatment of dry birth when the amniotic fluid had been prematurely released. A 2-foot (60 cm) section of bark was steeped in a gallon (4.5 l) of well water. One quart (1.1 l) of the extract was drunk.

The Cayuga aboriginal peoples used slippery elm, *U. rubra,* in the full management of childbirth. To facilitate the birthing process, the raw inner bark was chewed. Then, to help parturition, a decoction of bark chips of slippery elm, *U. rubra,* was mixed with bark chips of blue beech, *Carpinus caroliniana.* Blue beech is a member of the birch family and is getting quite rare. This was boiled in well water until the mucilage was extracted, the water becoming viscous. This was drunk. Then after the birthing process, bark chips of the American elm, *U. americana,* was added to the decoction and steeped. This water was drunk to prevent inflammation and to soothe torn membranous tissue.

The mucilage of slippery elm, *U. rubra,* was used as a poultice for sore eyes. Cold sores were also treated with a poultice by the Meskwaki aboriginal peoples. Sometimes these poultices were put on open wounds and also used to heal abscesses. The Chippewa peoples used the bark decoction for the treatment of ulcerated throats as a gargle. It was also used for infected kidneys. As much liquid as possible containing the mucilage extract was drunk.

The Cayuga peoples treated a form of sleeping paralysis with slippery elm. A black cherry, *Prunus serotina,* was selected with a black center. Four chips of 8 inches (20 cm) in length were taken from the heart of the tree. This was put into 6 quarts (5.7 l) of well water. The bark was stripped from the east side of the slippery elm, *U. rubra.* This was taken at 6 feet (1.8 m) from the ground. This mixture was boiled. Around 5 ounces (130 ml) dose was taken and repeated.

Ulmus glabra × *U. carpinifolia,* known as the Camperdownii elm, adding grace and magic in the design of a small garden

Surprise, surprise! The urban forest throws a sport of genetic resistance to Dutch elm disease. This tree is being cloned for future research.

The Chinese used the smooth-leafed elm, *U. campestris,* in the surface treatment of second-degree burns. Five parts of powdered endoderm were added to two parts Chinese philodendron, *Philodendron chinensis,* as a home remedy.

Slippery elm, *U. rubra,* is now used in lozenges for sore throats. It is also used in a malted form as a convalescent drink and it is used in the medical management of food poisoning and colitis as a mucosal anti-inflammatory.

ECOFUNCTION

The elm is a benign tree. It is found in great numbers throughout the north temperate forests of the world. It is a tree that guarantees migration food for many insect and bird populations as the warmth creeps up the face of the globe in spring. In North America elm flowers produce nectar and pollen for the insects. During the summer months, honey dew is also produced. In Europe pollen is produced and nectar, only occasionally, in warmer areas.

The young leaves feed porcupines and squirrel populations in the early spring. As the leaves expand, they act as feeding hosts to many species of common butterflies, the question mark, *Polygonia interrogationis;* the comma, *P. comma;* the mourning cloak, *Nymphalis antiopa;* and all over the western areas of North America, the Zephyr anglewing, *P. zephyrus.* These butterflies may use the nicotine alkaloid trigonelline from the elm for protection against predation in flight.

Elm seeds are an important food for songbirds, wood ducks, ring-necked pheasants, rodents, rabbits, and deer. The design of the elm is such that they make excellent perching trees. The birds can expose their feathers to facilitate the conversion of vitamin D. The birds then preen themselves and ingest this vitamin so essential for health and egg laying. The design also helps in the safe nesting of some songbirds like the northern orioles. The Baltimore oriole, *Icterus galbula,* in particular, likes to build cradle-like nests at the end of swinging, upturned branches of the American elm, *U. americana.*

The elm employs a very wise feeding strategy. All of the new season's growth on the tree, even the flowers and samaras, is covered with glandular hairs. These hairs are filled with at least two distinct compounds that taste salty, are nicotine-based, and are sexually stimulating. These taste enticers are the basis of the Western world fast food industry. It works for the masses and it works for the elm. The salty, stimulating, chemical is called caffearine or trigonelline. Some of this chemical is found in human urine after cigarette smoking. The sexually stimulating chemical is a glucocorticosteroid called dioscin. This famous chemical is also found in wild yams, *Dioscorea villosa* of the Dioscoreaceae family. Wild yam or rheumatism root has almost disappeared from North American habitats. This legendary yam has the same saponin glucocorticosteroid that also made the trillium, *Trillium erectum,* rhizomes indispensable as birthing agents for the aboriginal peoples. This chemical mix is found in all of the elms, but is found in quite large amounts mixed with the mucilage of the slippery elm, *U. rubra,* for some reason. It may also be found in the single surviving member of an entire genus of elm whose other members are extinct. This survivor is the plane tree or water elm, *Planera aquatica.*

A common tissue called cork is strangely found on some elms. It occurs on the trees as flanges. It is found on the rock elm, *U. thomasii,* and on the southern, winged elm, *U. alata,* which is also known as the cork elm or the *wahoo,* as named by the Cree. This cork contains about 60 percent suberin, which is, among other things, a sound attenuator. It could mean that these trees have evolved a different means of storing electrons from sunlight.

The mixture of glandular or domatal hairs and honey dew on the elm makes it an ideal microscopic habitat for predation and prey within the insect world. The predatory insects hosted by the elm help the entire forest habitat.

Shade is the trump card of the elm both as urban and actual forests. In particular the American elm, *U. americana,* acts like a giant umbrella, casting a globe of shade on the earth around the tree. This in turn forms a canopy for seedling trees and for wildflowers. This type of shade acts as a nursery ground for the development of seedling and sapling trees. Without this type of shade, there would be no forest. The introduced Chinese elm, *U. parvifolia,* also produces a fine shade for urban areas.

The elm is a very good antiturbulent agent in the forest. Despite its umbrella canopy, the elm is a very wind-stable tree and is rarely blown down in extreme gales. One species, the Wych elm, *U. glabra,* of Scotland is a superior tree for stability, especially in coastal areas. It is also excellent inland as a barrier for wind.

BIOPLAN

Elms should be protected and replanted. Resistant species should be sought out and be planted back into the deciduous forests of North America. The dominant American elms, *U. americana,* are disappearing from the rural landscape at an alarming rate. These species should be protected in private forests until they recover. The cycle for resistance emergence is about fifty to one hundred years. All of the elm species should be bioplanned back into rural areas.

Ulmus thomasii, rock elm, with its flare of cork flanges on the young limbs

Elm wood has some annoying characteristics to the wood splitter. The grain of the wood appears to be laid down in a cross-hatch pattern that makes it difficult to chop into firewood. But it is this wood morphology that makes it excellent in its green state to hold nails and bolts. This makes elm framing particularly earthquake proof as the wood fiber has an ability to expand and bend like an accordion. This aspect of the wood has made it an excellent choice for chairs. The original Windsor chairs were made from elm. Farming implements such as harrows were made from elm. In the ancient past, chariot wheels were made from elm and even door hinges.

Elm wood survives for long periods underground when not exposed to air. The water system of London was constructed from elm pipes that were revamped in 1930. The elm had with-

Chrysanthemum rubellum 'Mary Stoker'. The butter color of fall *Ulmus,* elm, species can be picked up and spun into a garden design using this heritage perennial chrysanthemum.

stood three hundred years of use underground and was dug up in almost perfect condition. Elm wood was also used in archery for longbows. The elm wood from the rock elm, *U. thomasii,* is particularly fine for hockey sticks, camogie sticks, or any sporting article that is required to withstand impact. Because of workability and availability, the ash has replaced the elm in many applications.

The slippery elm, *U. rubra,* which seems to be getting very scarce, could be grown for medicinal harvesting and for the spice industry.

The elm species could also be the basis of an important North American mushroom industry for the haute cuisine world trade. The various species of morels are rated second to truffles. Morels as a food item could be a cottage industry in North America, as truffles are in France. The morel mushroom is easily harvested and stored, and ships well, either fresh or dried. These, too, could be an excellent cash crop for the farming community if this industry were to be developed, much as the other mushroom industries are being developed and expanded.

DESIGN

The elms fit into the larger home landscape. They should be planted at some distance from the house for both perspective and to give the garden an appearance of depth and height. A number of hybrid elms are now on the market to replace the diseased American elm, *U. americana.* These have cultivar names such as 'Dynasty', 'Thompson', 'Jacan', 'Regal', 'Pioneer', and 'Homestead'. These cultivars are resistant to Dutch elm disease. In the United States, the *Zelkova serrata,* the Japanese keaki, is also being planted as a substitute for the elm. An elm hybrid called the 'Urban Elm' is becoming very popular in Europe but lacks the majesty of the native elm.

The slippery elm, *U. rubra,* is an ideal specimen tree for a large garden. Although the leaves and buds have a coarse texture, the tonal shades of the reddish glandular hairs add interest. This can be picked up in color by a mass planting of *Chrysanthemum rubellum* 'Mary Stoker'. This perennial chrysanthemum lasts for a long time in flower and adds sunshine and a certain crispness to the elm nearby.

Quintessentially North American is the combination of American elm, *U. americana,* with mass plantings of a white meadow of *Aster lateriflorus* 'Horizontalis'. This native perennial is the exact miniature of the American elm, laden down in the fall with myriads of tiny white daisies aligned in a horizontal plane. Both elm and aster can lay claim to drought and sunshine, both hallmarks of North American gardening.

As a specimen tree of great majesty, from its straight soaring trunk, to its winged branch flanges of cork, is the rock elm, *U. thomasii.* It is a tree of beauty and distinction and lends dignity to any garden.

Hardy and spectacular, certainly as it ages, is the top-grafted version of *U. glabra,* the Wych elm, called the Camperdownii elm. It is a small weeping elm that is hardy in zones 3–9. It is also disease resistant. It has fine ornamental value, especially when placed near the expanse of a water garden where it can be reflected.

Ulmus Species and Cultivars of Merit

SCIENTIFIC NAME	COMMON NAME	ZONES
Ulmus americana 'American Liberty'	Liberty elm	3–9
U. glabra	Wych elm (Scotch elm)	3–9
U. g. × *U. carpinifolia*	Camperdownii elm	3–9
U. × *hollandica* var. 'Vegeta' × *U. carpinifolia*	Urban elm	3–9
U. × *h.* var. 'Vegeta' × *U. c.* 'Dynasty'	Dynasty elm	3–9
U. × *h.* var. 'Vegeta' × *U. c.* 'Homestead'	Homestead elm	3–9
U. × *h.* var. 'Vegeta' × *U. c.* 'Jacan'	Jacan elm	3–9
U. × *h.* var. 'Vegeta' × *U. c.* 'Pioneer'	Pioneer elm	3–9
U. × *h.* var. 'Vegeta' × *U. c.* 'Regal'	Regal elm	3–9
U. × *h.* var. 'Vegeta' × *U. c.* 'Thompson'	Thompson elm	3–9
U. rubra	Slippery elm	3–9
U. serotina	September elm	3–10
U. thomasii	Rock elm	3–9

The mature bark of *Ulmus americana*, American elm, a sight seldom seen nowadays

References

Baldwin, I. T. "Chemical SOS Not Just for Farm Lab Plants." *Science News,* March 2001, 166.

Beresford-Kroeger, Diana. "King of the Forest." *Nature Canada,* spring 2000, 16–19.

———. "A Summer Beauty." *Nature Canada,* winter 1999, 18–19.

———. *Bioplanning a North Temperate Garden.* Kingston, Ont.: Quarry Press, 1999.

Boon, Heather, and Michael Smith. *The Botanical Pharmacy.* Kingston, Ont.: Quarry Press, 1999.

Borror, Donald J., and Richard E. White. *A Field Guide to the Insects of America North of Mexico.* Boston: Houghton Mifflin, 1970.

Brodo, Irwin M., Sylvia Duran Sharnoff, and Stephen Sharnoff. *Lichens of North America.* New Haven: Yale University Press, 2001.

Budavari, S. *The Merck Index: An Encyclopedia of Chemicals, Drugs, and Biologicals.* 11th ed. Rahway, N.J.: Merck, 1989.

Casselman, Bill. *Canadian Garden Words.* Toronto: Little Brown, 1997.

Chenoweth, Bob. "The History, Use, and Unrealized Potential of a Unique American Renewable Nature Resource." *Northern Nut Growers Association Annual Report* 86 (1995): 18–20.

Clausen, Ruth Rogers, and Nicholas H. Ekstrom. *Perennials for American Gardens.* New York: Random House, 1989.

Cody, J. *Ferns of the Ottawa District.* Ottawa: Canada Department of Agriculture, 1956.

Collingwood, G. H., and Warren D. Bush. *Knowing Your Trees.* Washington, D.C.: American Forestry Association, 1974.

Cormack, R. G. H. *Wild Flowers of Alberta.* Edmonton: Queen's Printers, 1967.

Davies, Karl M., Jr. "Some Ecological Aspects of Northeastern American Indian Agroforestry Practises." *Northern Nut Growers' Association Annual Report* 85 (1994): 25–39.

Densmore, F. *Indian Use of Wild Plants for Crafts, Food, Medicine, and Charms.* Ohsweken: Iroqrafts, 1993.

Fell, Barry. *Bronze Age America.* Toronto: Little Brown, 1982.

Flint, Harrison L. *Landscape Plants for Eastern North America.* New York: John Wiley and Sons, 1983.

Fox, Katsitsionni, and Margaret George. *Traditional Medicines.* Cornwall: Mohawk Council of Akwesasne, 1998.

Frazer, James G. *The Golden Bough.* New York: Avenel Books, 1981.

Granke, L. J. "Genetic Resources of Carya in Vietnam and China." *Northern Nut Growers' Association Annual Report* 82 (1991): 80–87.

Hamilton, J. W. "Arsenic Pollution Disrupts Hormones." *Science News,* March 2001, 164.

Heatherley, Ana Nez. *Healing Plants: A Medical Guide to Native North American Plants and Herbs.* New York: Lyons Press, 1998.

Herity, Michael, and George Eogan. *Ireland in Prehistory.* New York: Routledge, 1996.

Herrick, James W. *Iroquois Medical Botany.* Syracuse: Syracuse University Press, 1995.

Hillier, Harold. *The Hillier Manual of Trees and Shrubs.* Newton Abbot: David and Charles Redwood, 1992.

Hosie, R. C. *Native Trees of Canada.* Ottawa: Department of Fisheries and Forestry, 1969.

Howes, F. N. *Nuts: Their Production and Everyday Use.* London: Faber and Faber, 1948.

———. *Plants and Beekeeping.* London: Faber and Faber, 1979.

Jaynes, Richard A. *Nut Tree Culture in North America.* Hamden, Conn.: Northern Nut Growers' Association, 1979.

Kingsbury, John M. *Poisonous Plants of the United States and Canada.* Englewood Cliffs, N.J.: Prentice-Hall, 1964.

Klots, Alexander B. *A Field Guide to Butterflies of North America, East of the Great Plains.* Boston: Houghton Mifflin, 1951.

Krieger, Louis C. *The Mushroom Handbook.* New York: Dover, 1967.

Krochmal, Arnold, and Connie Krochmal. *The Complete Illustrated Book of Dyes from Natural Sources.* New York: Doubleday, 1974.

Lee, Robert Edward. *Phycology.* 2d ed. Cambridge: Cambridge University Press, 1995.

Lellinger, David B. *A Field Manual of Ferns and Fern Allies of the*

United States and Canada. Washington, D.C.: Smithsonian Institution Press, 1985.
Lewis, Walter H., Memory P. F. Elvin-Lewis. *Medical Botany: Plants Affecting Man's Health.* Toronto: John Wiley and Sons, 1979.
Liberty Hyde Bailey Hortorium. *Hortus Third: A Concise Dictionary of Plants Cultivated in the Unites States and Canada.* New York: Macmillan, 1976.
Little, Elbert L. *Trees.* New York: Alfred A. Knopf, 1980.
Megan, Ruth, and Vincent Megan. *Celtic Art.* London: Thames and Hudson, 1999.
Myers, Norman. *Gaia: An Atlas of Planet Management.* New York: Doubleday, 1984.
Mullins, E. J., and T. S. McKnight. *Canadian Woods: Their Properties and Uses.* Toronto: University of Toronto Press, 1981.
Peterson, Roger Tory, and Margaret McKenny. *A Field Guide to Wild Flowers of Northeastern and North-central North America.* Boston: Houghton Mifflin, 1968.
Phillips, Roger, and Martyn Rix. *Perennials.* 2 vols. New York: Random House, 1991.
Pirone, P. P. *Tree Maintenance.* 6th ed. Oxford: Oxford University Press, 1988.
Rackham, Oliver. *The Illustrated History of the Countryside.* London: George Weidenfeld and Nicolson, 1994.
Rupp, Rebeca. *Red Oaks and Black Birches: The Science and Lore of Trees.* Pownal, Vt.: Storey Communications, 1995.
Schopmeyer, C. S. *Seeds of Woody Plants in the United States.* Washington, D.C.: Forest Service, U.S. Department of Agriculture, 1974.
Small, Ernest, and Paul M. Caitling. *Canadian Medical Crops.* Ottawa: National Research Council of Canada, 1999.
Smith, Russell J. *Tree Crops: A Permanent Agriculture.* New York: Devin-Adair, 1953.
Stuart, Malcolm. *The Encyclopedia of Herbs and Herbalism.* London: Orbix, 1979.
Taylor, Kathryn S., and Stephen F. Hamblin. *Handbook of Wildflower Cultivation.* New York: Macmillan, 1963.
Uhari, Matti K. "A Sugar Averts Some Ear Infections." *Science News,* October 1998, 287.
Waldron, G. E. *The Tree Book: Tree Species and Restoration Guide for the Windsor-Essex Region.* Windsor, Ont.: Project Green, 1997.
Wickens, G. E. *Edible Nuts.* Rome: Food and Agriculture Organization of the United Nations, 1995.

Photography

Photographs for this text were taken with a Canon TL-QL 35mm SLR camera using the following Canon lenses: FD 70-210 mm zoom F/4, FL 50 mm F/1.8, and FD 28 mm F/2. For virtually all the photographs, Kodachrome 64 or Kodachrome 200 color transparency film was used. All photographs were taken in natural light without enhancement. The photographer wishes to thank Moyra V. Crich of T.J. Photo Ltd., Ottawa, Ontario, for photographic advice and support and the staff of the Canadian Camera Service Centre Inc., Ottawa, Ontario, for cheerfully maintaining his ancient equipment.

Index